I. S. Iohvidov, M. G. Krein, H. Langer

Introduction to the Spectral Theory of Operators in Spaces with an Indefinite Metric

Mathematical Research · Mathematische Forschung

Wissenschaftliche Beiträge
herausgegeben von der
Akademie der Wissenschaften der DDR
Institut für Mathematik

Band 9

Introduction to the Spectral Theory of Operators in Spaces with an Indefinite Metric

by
I. S. Iohvidov
M. G. Krein
H. Langer

Introduction to the Spectral Theory of Operators in Spaces with an Indefinite Metric

by
I. S. Iohvidov
M. G. Krein
H. Langer

Akademie-Verlag · Berlin 1982

Autoren:

Prof. Dr. I. S. Iohvidov
Voronesh State University, Voronesh, U.S.S.R.

Prof. Dr. M. G. Krein
Southern Centre of the Academie of Science, Odessa, U.S.S.R.

Prof. Dr. H. Langer
Technical University, Dresden, GDR

Die Titel dieser Schriftenreihe werden vom Originalmanuskript
der Autoren reproduziert.

Erschienen im Akademie-Verlag, DDR-1086 Berlin, Leipziger Str. 3—4
Lektor: Dr. Reinhard Höppner

Lizenznummer: 202 · 100/412/82
Gesamtherstellung: VEB Kongreß- und Werbedruck, 9273 Oberlungwitz
Umschlag: Rolf Kunze
Bestellnummer: 763 041 3 · (2182/9) · LSV 1035
Printed in GDR
DDR 15,— M

TABLE OF CONTENTS

INTRODUCTION

This paper was written on the basis of the article [19a] of I.S. Iohvidov and M.G. Kreĭn, which first appeared in 1956. At that time, about one dozen articles on spaces with indefinite metric and their linear operators had been published. Now, however, more than 300 papers on this subject have appeared: see the bibliography of the survey article [4c]. Also, this theory has gone far beyond the framework of the spaces $\Pi_\varkappa$, and in 1974 the monograph [9] appeared, offering an excellent presentation of a series of questions in this theory. This monograph, however, is written from a more general point of view, and spaces $\Pi_\varkappa$, which are the simplest (infinite dimensional) examples of spaces with indefinite metric, are treated there only in the last chapter. On the other hand, during the last 20 years the deepest and most complete results have been obtained for spaces $\Pi_\varkappa$. In this connection we mention the spectral decomposition for selfadjoint and unitary operators [24a], the theory of extensions and generalized resolvents of Hermitian operators [24d], generalizations of the classical theorems of Bochner and Stone [21e], [32], the theory of commutative operators algebras and the representations of algebras and groups in $\Pi_\varkappa$ (a fairly complete survey is given in [33]), and conditions for the completeness of root vectors of selfadjoint and dissipative operators [18c], [4b], [3b]. Moreover, selfadjoint and unitary operators as well as some new classes of operators in spaces $\Pi_\varkappa$ have found important applications, see e.g. [21c], [24c,e,f], [29], [1]. Therefore, it seems desirable to us to have a new introduction to the theory of spaces $\Pi_\varkappa$ available.

With this aim in mind, the present paper has been prepared. As [19a], it still has an introductory character and contains neither the new results nor the applications mentioned above. Moreover, also its basic skeleton is the same as in [19a]. However, some old results are presented in a new way, and these new methods sometimes completed the results so essentially that, in fact, they grew into new results with new corollaries (see, e.g., Theorems 11.2 and 11.3). Also some incorrect statements in [19a], which were indicated by its authors first in Uspehi Math. Nauk 11 (1956), n° 4 (70), then in Trudy Moskov. Mat. Obšč. 6 (1957) and, finally, in [19b], as well as some other minor defects, have been corrected. This was the joint work of all three authors.

The main alterations compared with the article [19a] are the following. In the first two sections characterisations of completeness and norm convergence in $\Pi_{\varkappa}$ have been added, which are useful for applications. The § 6 is considerably extended by new classes of operators; the § 8, dealing with the Cayley transformation, has been revised. The part of § 9, concerning extensions of Hermitian operators, and § 10, which now offers the definition of the angle operator, have been completely revised. The angle operator, introduced independently by R.S. Phillips [34] and Ju.P. Ginzburg [16b], is an important tool in the theory of spaces with indefinite metric and its applications.

The § 11, which contains the most interesting results of this article, has been written anew. It now includes a series of new statements, and some results of [19a] have been revised and completed. In [19a] the Riesz projectors appeared only in the Supplement. Now they play an essential role in § 11. Of special importance in § 11 is the notion of a uniformly expansive operator. Moreover, in this section the reader can find a fairly detailed study of expansive, contractive and dissipative operators.

The list of references is restricted to those quoted in the text. The reader can find a fairly complete bibliography on spaces with indefinite metric and their linear operators in [4c] (see also [9]).

CHAPTER 1

LINEAR SPACES $\Pi_\varkappa$ WITH INDEFINITE METRIC

§ 1. Axiomatics of the spaces $\Pi_\varkappa$

1.1. We define a space $\Pi_\varkappa$ ($\varkappa$ a positive integer) by means of a system of axioms I - VI (the precise definition of some of the terms encountered in these axioms is given in 1.2).

I. $\Pi_\varkappa$ is a linear space (with multiplication of vectors by scalars from the field of complex numbers).

The zero vector of the space $\Pi_\varkappa$ will be denoted by θ.

II. There is given in $\Pi_\varkappa$ a bilinear metric, i.e., for any two vectors $x,y \in \Pi_\varkappa$ there is defined a complex number (x,y), their scalar product, which has the properties:
a) $(y,x) = \overline{(x,y)}$, b) $(\lambda x + \mu y, z) = \lambda(x,z) + \mu(y,z)$, where $x,y,z \in \Pi_\varkappa$, and λ,μ are any complex numbers.

III. $\Pi_\varkappa$ contains no vector other than θ which is orthogonal in the sense of this scalar product to the whole of $\Pi_\varkappa$.

IV. $\Pi_\varkappa$ contains at least one subspace P of dimension $\varkappa$ on which the form (x,x) $(x \neq \theta)$ is positive (a "positive" subspace).

V. For any $(\varkappa+1)$ vectors $x_1,x_2,\ldots,x_\varkappa,x_{\varkappa+1}$ in $\Pi_\varkappa$ the Hermitian form $\sum_{j,k=1}^{\varkappa+1} (x_j,x_k)\,\xi_j\bar{\xi}_k$ contains not more than $\varkappa$ positive squares.

VI. Amongst all the decompositions of $\Pi_\varkappa$ into a direct orthogonal sum

$$\Pi_\varkappa = \Pi_+ \oplus \Pi_- , \qquad (1.1)$$

where Π_+ is a $\varkappa$-dimensional "positive" subspace and Π_- its orthogonal complement in $\Pi_\varkappa$, which exist in view of the preceding axioms, there is at least one with the property that, on introducing the norm $|x| := \sqrt{-(x,x)}$ $(x \in \Pi_-)$, Π_- becomes a complete

Banach space.[1])

Note. It is easy to see that the axiomatics of the space $\Pi_\varkappa$ can be simplified in the following way:
Axioms I and II remain unchanged, but instead of the other four axioms (III - VI) we use the following:

> The space $\Pi_\varkappa$ admits a decomposition into an orthogonal (with respect to the given scalar product) direct sum $\Pi_\varkappa = \Pi_+ \oplus \Pi_-$[2]), where Π_+ is a $\varkappa$-dimensional Hilbert space with scalar product (x,y), and Π_- is a Hilbert space with scalar product $[x,y] = -(x,y)$.

If in this axiom we do not require that the space Π_+ is of dimension $\varkappa$, allowing it to be an arbitrary (in general infinite-dimensional) Hilbert space with scalar product (x,y), we simply write Π instead of $\Pi_\varkappa$ and call this space Π a space with regular indefinite metric.[3]) The theory of linear operators in these spaces is more difficult and beyond the scope of this article. We gave the definition of the space Π only in order to simplify some notes and remarks.

1.2. We shall now make the terminology used in the formulation of axioms I - VI more precise, and at the same time we shall establish some results which follow from these axioms. Note that unlike to the ordinary scalar product ("definite metric") that introduced in axiom II is not positive definite, i.e. there are three possibilities,

$$(x,x) > 0, = 0 \text{ or } < 0 \quad (x \in \Pi_\varkappa) .$$

Definition 1.2. A subset $M \subset \Pi_\varkappa$ is called positive (negative, neutral, nonnegative, etc.) if all its nonzero elements are positive (negative, neutral, nonnegative, etc., respectively).

Positive and negative (nonnegative and nonpositive) subsets will be named by the common term definite (semidefinite, respectively) subsets.

---

1) We do not exclude the case in which Π_- is finite-dimensional. In that case, $\Pi_\varkappa$ is also finite-dimensional.

2) We denote direct sums as usual by the symbol $\dot{+}$. The symbol $\oplus$ here and in what follows denotes a direct sum whose components are orthogonal to one another.

3) In [9] these spaces are called Krein spaces.

Using the terminology introduced in Definition 1.2, our axiom V may be equivalently expressed as follows:

> $\Pi_\varkappa$ does not contain a positive subspace of dimension $\varkappa + 1$ (cf. Lemma 1.2, below).

Definition 1.3. Vectors $x,y \in \Pi_\varkappa$ are called orthogonal if $(x,y) = 0$. Orthogonality of a vector to a set and orthogonality of two sets in $\Pi_\varkappa$ are defined correspondingly.

Clearly a neutral vector is orthogonal to itself. For brevity orthogonality of vectors and sets in $\Pi_\varkappa$ will be denoted by the symbol $\perp$.

Definition 1.4. The orthogonal complement of a set M in $\Pi_\varkappa$ is the set of all vectors in $\Pi_\varkappa$ which are orthogonal to M; we denote it by $M^\perp$.

We call a set $M \subset \Pi_\varkappa$ a linear manifold if for any two vectors $x,y \in M$, and any complex numbers λ and μ we have $\lambda x + \mu y \in M$. A finite-dimensional subspace of $\Pi_\varkappa$ is a particular case of a linear manifold. The concept of an infinite-dimensional subspace will be introduced below (cf. 3.1). It is easy to see that a linear manifold which is composed entirely of neutral vectors is orthogonal to itself, i.e., the scalar product is identically zero on it.

Lemma 1.1. If P is a positive $\varkappa$-dimensional subspace of the space $\Pi_\varkappa$ (see axiom IV) and N is its orthogonal complement, then N is a negative linear manifold, and $\Pi_\varkappa$ is the direct sum $\Pi_\varkappa = P \oplus N$.

Proof. Since P is an ordinary unitary space, one can find an orthonormal basis for it,

$$e_1,\ldots,e_k, \quad (e_j,e_k) = \delta_{jk} \quad (j,k = 1,\ldots,\varkappa) .$$

Clearly N is the set of all vectors $y \in \Pi_\varkappa$ which satisfy

$$(y,e_j) = 0 \quad (j = 1,\ldots,\varkappa) .$$

From this, using axiom II, b), we deduce that N is a linear manifold. For any $x \in \Pi_\varkappa$ we can find a unique set of complex numbers $\xi_1,\xi_2,\ldots,\xi_\varkappa$ such that the vector

$$z = x - \xi_1 e_1 - \xi_2 e_2 - \ldots - \xi_\varkappa e_\varkappa$$

lies in N.

To prove this it is enough to take $\xi_j = (x,e_j) \quad (j = 1,\ldots,\varkappa)$. Then $z \in N$, since

$$(z,e_k) = (x,e_k) - \sum_{j=1}^{\varkappa} \xi_j(e_j,e_k) = (x,e_k) - \xi_k = 0 \quad (k = 1,\dots,\varkappa) .$$

If now we have also that $x = \sum_{j=1}^{\varkappa} \eta_j e_j + z'$, where $z' \in N$, then

$$\sum_{j=1}^{\varkappa} (\xi_j - \eta_j)e_j + (z - z') = \theta \quad (z - z' \in N) , \tag{1.2}$$

and on forming the scalar products of both sides of (1.2) with the e_k $(k = 1,\dots,\varkappa)$ successively, we see that $\xi_k = \eta_k$ $(k = 1,\dots,\varkappa)$ and $z = z'$.

Thus, as $\xi_1 e_1 + \xi_2 e_2 + \dots + \xi_\varkappa e_\varkappa \in P$ we have established the decomposition $\Pi_\varkappa = P \oplus N$. It remains to show that the linear manifold N is negative.

Let y be any vector in N. We consider the hermitian form of order $\varkappa + 1$:

$$\sum_{j,k=1}^{\varkappa} (e_j,e_k)\xi_j\bar{\xi}_k + (y,y)|\xi_{k+1}|^2 = \sum_{j=1}^{\varkappa} |\xi_j|^2 + (y,y)|\xi_{k+1}|^2 .$$

It follows from axiom V that $(y,y) \leq 0$. Thus the scalar square (x,x) is nonpositive on the linear manifold N.[1] But then for any two vectors $x,z \in N$ we have the Cauchy-Schwarz-Bunjakovskiĭ inequality

$$|(y,z)|^2 \leq (y,y) \cdot (z,z) . \tag{1.3}$$

Now let $y \in N$ and $y \neq \theta$. If $(y,y) = 0$, then $(y,z) = 0$ for any $z \in N$ by (1.3). Thus we should have $y \perp N$, $y \perp P$ and so $y \perp \Pi_\varkappa$, which contradicts axiom III. Lemma 1.1 is thus completely proved.

<u>1.3.</u> Axiom IV established the existence of a $\varkappa$-dimensional positive subspace $P \subset \Pi_\varkappa$. The lemma that we have just proved with the help of axiom V enables us to see that the dimension $\varkappa$ is the maximum of the dimensions possible for positive (and nonnegative) linear manifolds of the space $\Pi_\varkappa$.

<u>Lemma 1.2.</u> If L is any nonnegative linear manifold in $\Pi_\varkappa$, then the dimension of L is not greater than $\varkappa$.

<u>Proof.</u> Let P be a positive $\varkappa$-dimensional subspace of $\Pi_\varkappa$ and $\{e_j\}_{j=1}^{\varkappa}$ an orthonormal basis of it. By Lemma 1.1

---

[1]) We note that this assertion is proved without using axiom II. Moreover, the complete proof of Lemma 1.1 nowhere uses axiom VI.

$\Pi_\varkappa = P \oplus N$,

where N is a negative linear manifold. The vectors $x, y \in L$ have unique representations

$$\begin{aligned} x &= \xi_1 e_1 + \xi_2 e_2 + \ldots + \xi_\varkappa e_\varkappa + u \quad (u \in N) , \\ y &= \eta_1 e_1 + \eta_2 e_2 + \ldots + \eta_\varkappa e_\varkappa + v \quad (v \in N) . \end{aligned} \tag{1.4}$$

Let us suppose that $\xi_j = \eta_j$ $(j = 1,\ldots,\varkappa)$. Then it follows from (1.4) that

$$x - y = u - v .$$

But $x - y \in L$, i.e., $(x - y, x - y) \geqq 0$, while $u - v \in N$, i.e., $(u - v, u - v) \leqq 0$. Consequently, $u = v$ and $x = y$. Thus, any vector x in L is completely determined by its coordinates $\xi_1,\ldots,\xi_\varkappa$, i.e., the dimension of L is not greater than $\varkappa$.

1.4. Now we proceed to clarify axiom VI. We note that, apart from the $\varkappa$-dimensional positive subspace P, the existence of which is required by axiom VI, $\Pi_\varkappa$ can contain other positive subspaces, the dimensions of which by Lemma 1.2 cannot exceed $\varkappa$.

If the dimension of a positive subspace $P' \subset \Pi_\varkappa$ is equal to $\varkappa$, then, apart from the decomposition $\Pi_\varkappa = P \oplus N$, we have, evidently, also a decomposition $\Pi_\varkappa = P' \oplus N'$, where once again N' is a negative linear manifold, and so on.

It is easy to verify that any of the linear manifolds N,N', etc. can be made into a normed linear space, with the norm

$$|x| = \sqrt{-(x,x)} \quad (x \in N) .$$

Here it is not known whether the space N is complete with respect to this norm. It appears that in order to ensure that this property should hold for all N,N', etc., it is enough to introduce axiom VI, requiring that it be fulfilled for at least one of the decompositions of this sort (cf., Theorem 2.1 below).

1.5. The space $\Pi_\varkappa$ can also be defined by the axioms I - V above and the following axiom VI', which does not make use of a special decomposition (1.1). In order to formulate it we introduce some notation. A sequence (z_n) of elements of the space $\Pi_\varkappa$ satisfying axioms I - V, is called weakly convergent, if there exists a $z_0 \in \Pi_\varkappa$ such that $(z_n - z_0, y) \to 0$ for $n \to \infty$ and all $y \in \Pi_\varkappa$, and a weak Cauchy sequence if for each $y \in \Pi_\varkappa$ the numbers (z_n, y), $n = 1,2,\ldots,$ form a Cauchy sequence.

VI'. The space $\Pi_\varkappa$ is weakly sequentially complete, that is, each weak Cauchy sequence is weakly convergent.

Indeed, if VI holds and (x_n) is a weak Cauchy sequence: $(x_n - x_m, y) \to 0$, $n,m \to \infty$, for all $y \in \Pi_\varkappa$, we choose an orthonormal basis $e_1, e_2, \ldots, e_\varkappa$ in Π_+ and consider the representation

$$x_n = \xi_{n1}e_1 + \ldots + \xi_{n\varkappa}e_\varkappa + x_n^-, \quad x_n^- \in \Pi_- .$$

Taking $y = e_j$, it follows $\xi_{nj} - \xi_{mj} \to 0$, $n,m \to \infty$, $j = 1,\ldots,\varkappa$, hence there exist complex numbers ξ_j such that $\xi_{nj} \to \xi_j$ if $n \to \infty$, $j = 1,\ldots,\varkappa$. Moreover, $(x_n^- - x_m^-, y^-) \to 0$ if $n,m \to \infty$, for all $y^- \in \Pi_-$. But the (Hilbert)space Π_- is weakly sequentially complete, hence there exists an $x_0^- \in \Pi_-$ such that $(x_n^- - x_0^-, y^-) \to 0$ if $n \to \infty$ and for all $y^- \in \Pi_-$. If $x_0 = \sum_{j=1}^{\varkappa} \xi_j e_j + x_0^-$, we have $(x_n - x_0, y) \to 0$, $n \to \infty$, for all $y \in \Pi_\varkappa$.

Suppose now, conversely, that VI' holds and a decomposition $\Pi_\varkappa = \Pi_+ \oplus \Pi_-$ according to Lemma 1.1 is given. Let (x_n^-) be a Cauchy sequence in Π_- with respect to the norm $|\cdot|$. Then we have $(x_n^- - x_m^-, y^-) \to 0$, $n,m \to \infty$, for all $y^- \in \Pi_-$, and $(x_n^- - x_m^-, y) \to 0$, $n,m \to \infty$, for all $y \in \Pi_\varkappa$. By VI' there is an element $x_0 \in \Pi_\varkappa$ such that $(x_n^- - x_0, y) \to 0$, $n \to \infty$, for all $y \in \Pi_\varkappa$. As $(x_0, y) = \lim_{n\to\infty}(x_n^-, y)$ for all $y \in \Pi_\varkappa$, it follows that $x_0 \in \Pi_-$. Moreover,

$$|x_n^- - x_0|^2 = -\lim_{m\to\infty}(x_n^- - x_0, x_n^- - x_m^-) \leq |x_n^- - x_0| \lim_{m\to\infty}|x_n^- - x_m^-|$$

and $|x_n^- - x_0| \leq \lim_{m\to\infty}|x_n^- - x_m^-| \to 0$, $n \to \infty$; that is, Π_- is complete.

1.6. In concluding this section we note that we can also consider real spaces $\Pi_\varkappa$. The axioms I – VI need only unimportant modifications, which are as follows: In axiom I the field of scalars must be taken to be the real field. In axiom II the scalar product must be required to be real, $(x,y) = (y,x)$. Finally in axiom V the Hermitian form must be replaced by the corresponding quadratic form.

With these modifications all the Definitions 1.1 – 4 retain their former sense, and the Lemmas 1.1 and 1.2 remain valid. The same can be said of the Theorems 2.1 – 3.4 which follow, and which describe the geometric structure of the spaces $\Pi_\varkappa$.

§ 2. Topological properties of the spaces $\Pi_\varkappa$

2.1. First of all let us observe that with decomposition (1.1) we can introduce a new scalar product $[x,y]$ $(x,y \in \Pi_\varkappa)$ in $\Pi_\varkappa$, which is positive definite, i.e., $[x,x] > 0$ if $x \neq \theta$. To this end we represent each vector $x \in \Pi_\varkappa$ in the form

$$x = x_+ + x_- \quad (x_+ \in \Pi_+;\ x_- \in \Pi_-) .$$

Now, for any $x,y \in \Pi_\varkappa$ we put

$$[x,y] = (x_+,y_+) - (x_-,y_-) . \tag{2.1}$$

It is not difficult to see that $[x,y]$ has all the usual properties of a scalar product, including the positive definiteness. Obviously the subspaces Π_+ and Π_- are orthogonal not only with respect to the scalar product (x,y), but also with respect to the scalar product $[x,y]$. We see without difficulty that

$$[x,y] = (x,y) - 2(x_-,y_-) = -(x,y) + 2(x_+,y_+) . \tag{2.2}$$

Thus $\Pi_\varkappa$ can also be regarded as an ordinary unitary space H with the scalar product $[x,y]$, which we shall from now on for shortness call the <u>H-scalar product</u>, and with the norm

$$\|x\| = \sqrt{[x,x]} = \sqrt{(x_+,x_+) - (x_-,x_-)} . \tag{2.3}$$

From (2.1) we easily get

$$(x,y) = [x_+,y_+] - [x_-,y_-] \quad (x,y \in \Pi_\varkappa) .$$

This relation implies that the scalar product (x,y) $(x,y \in \Pi_\varkappa)$ is continuous with respect to the norm (2.3) in both variables x and y. More exactly, we have

$$|(x,y)| \leqq \|x\|\|y\| \quad (x,y \in \Pi_\varkappa) . \tag{2.4}$$

Indeed,

$$|(x,y)|^2 \leqq |[x_+,y_+] - [x_-,y_-]|^2 \leqq (\|x_+\|\|y_+\| + \|x_-\|\|y_-\|)^2 \leqq$$
$$\leqq (\|x_+\|^2 + \|x_-\|^2)(\|y_+\|^2 + \|y_-\|^2) = \|x\|^2\|y\|^2 .$$

For the vectors $x \in \Pi_-$ the norm (2.3) coincides with the norm $|x| = \sqrt{-(x,x)}$. Consequently, the subspace Π_-, which is complete with respect to the norm $|x|$ by axiom VI (§ 1), is also complete with respect to the norm $\|x\|$ $(x \in \Pi_-)$.

<u>Definition 2.1.</u> If L and M $(L \subset M)$ are two linear manifolds in $\Pi_\varkappa$, then the <u>dimension of M modulo L</u> means the dimension[1]) dim M/L of the linear manifold M/L[2]).

If dim M/L is finite, then one can easily see that in order that dim M/L = r it is necessary and sufficient that $M = L \dotplus N$, where dim N = r (see, e.g., [30]). Since $\Pi_\varkappa = \Pi_+ \oplus \Pi_-$, it follows that $\dim \Pi_\varkappa/\Pi_- = \varkappa < \infty$. In view of this the completeness of Π_- with respect to the norm $\|x\|$ $(x \in \Pi_-)$ implies the completeness of $\Pi_\varkappa$ with respect to the norm $\|x\|$ $(x \in \Pi_\varkappa)$.

<u>2.2.</u> In this way axiom VI allows us to regard the space $\Pi_\varkappa$ henceforth as a complete Banach space with the norm $\|x\|$ (see (2.3)).

<u>Theorem 2.1.</u> Let P be any $\varkappa$-dimensional positive subspace of $\Pi_\varkappa$. Then its orthogonal complement N is a complete space with respect to the norm

$$|x| = \sqrt{-(x,x)} \quad (x \in N) .$$

<u>Proof.</u> Consider the intersection of N and Π_-: $F = N \cap \Pi_-$. If $\{e_j\}_{j=1}^{\varkappa}$ is an orthonormal basis of Π_+, then clearly F is the set of all vectors f in N which satisfy the conditions $(f,e_j) = 0$ $(j = 1,\dots,\varkappa)$ or the equivalent conditions $[f,e_j] = 0$ $(j = 1,\dots,\varkappa)$. The equivalence follows from the fact that in general for any $x \in \Pi_\varkappa$

$$[x,e_j] = (x_+,e_j) - (x_-,\theta) = (x_+ + x_-,e_j) = (x,e_j) \quad (j = 1,\dots,\varkappa) .$$

It is easy to see that F is a linear manifold.

As in every unitary space, the H-scalar product $[x,y]$ $(x,y \in \Pi_\varkappa)$ is

---

1) We recall that in general the dimension (number of dimensions) of a Banach space or any of its linear manifolds L denotes [23] the least of the powers u_M of the sets M $(\subset L)$ whose linear spans are dense in L, i.e., the closure $\bar{M} \supset L$. Thus $\dim L = \min u_M$ is a cardinal number. If dim L is finite, then it coincides with the maximum number of linearly independent elements of L.

2) The linear manifold M/L is defined in the way that is usual for the factor group, i.e., as the collection of classes X of elements "equal modulo L": $x_1, x_2 \in X$ if $x_1 - x_2 \in L$. The class L is the zero element of M/L, the norm in M/L is defined by $\|X\| = \inf_{x \in X} \|x\|$.

continuous in both arguments relatively to the norm $\|x\|$ of (2.3). From this it follows that F is closed with respect to the norm $\|x\|$. But for $x \in F$ we have $\|x\| = |x|$, i.e., F is closed also with respect to the norm $|x|$.

Consequently, if $F = N$, the completeness of N with respect to the norm $|x|$ is proved.

The case when $F \subset N$, $F \neq N$ must now be considered. As N is a unitary (negative) space, we have $N = F \oplus E$, where E is the orthogonal complement of F in N. We show that the dimension of E (i.e., dim N/F) does not exceed $\varkappa$.

In fact, if we found linearly independent vectors $y_1, \ldots, y_{\varkappa+1} \in E$, then it would be possible to choose a nonzero system of constants $\xi_1, \ldots, \xi_{\varkappa+1}$ so that $\sum_{j=1}^{\varkappa+1} \xi_j y_j \in F$. For this it is sufficient to solve the homogeneous system

$$\sum_{j=1}^{\varkappa+1} \xi_j (y_j, e_k) = 0 \quad (k = 1, \ldots, \varkappa)$$

of $\varkappa$ linear equations with $(\varkappa + 1)$ unknowns $\xi_1, \ldots, \xi_{\varkappa+1}$. But such a system has a nontrivial solution. Hence

$\sum_{j=1}^{\varkappa+1} \xi_j y_j \in F$ and $\sum_{j=1}^{\varkappa+1} \xi_j y_j \in E$, therefore

$$\xi_1 y_1 + \ldots + \xi_{\varkappa+1} y_{\varkappa+1} = \theta ,$$

which contradicts the linear independence of the vectors $y_1, \ldots, y_{\varkappa+1}$.

Thus dim $N/F \leq \varkappa$. As it has been shown above that F is closed with respect to the norm $|x|$, N is complete with respect to $|x|$.

As a consequence of Theorem 2.1 we have a somewhat more general proposition.

Theorem 2.2. Let P and N be the subspaces of $\Pi_\varkappa$ which are defined in the statement of Theorem 2.1. Then the norms $\|x\|$ and $|x|$ are equivalent on N.

Proof. Since $[x,x] = -(x,x) + 2(x_+, x_+)$ for all $x \in \Pi_\varkappa$ (see (2.2)), it follows that $[x,x] \geq -(x,x)$, i.e. $\|x\| \geq |x|$ $(x \in N)$. Hence the identity mapping from N, equipped with norm $\|x\|$, onto N, equipped with norm $|x|$, is continuous, and a theorem of Banach [20] implies the continuity of the inverse mapping. This means the existence of a constant $c > 0$ such that $\|x\| \leq c|x|$ $(x \in N)$, and the theorem is proved.

Theorem 2.3. Let $\Pi_\varkappa = P \oplus N$, where P is a positive $\varkappa$-dimensional subspace. If in $\Pi_\varkappa$ we introduce the norm

$$x = \sqrt{(x^P,x^P) - (x^N,x^N)} \quad (x \in \Pi_\varkappa;\ x = x^P + x^N;\ x^P \in P;\ x^N \in N), \tag{2.5}$$

then it is equivalent to the norm $\|x\|$, defined by the decomposition $\Pi_\varkappa = \Pi_+ \oplus \Pi_-$ according to the formula (2.3).

Proof. Because of Theorem 2.2 the norms $\|x\|$ and $|x|$ are topologically equivalent on the subspaces N and Π_-. But $\dim \Pi_\varkappa/\Pi_- = \dim \Pi_\varkappa/N = \varkappa$, whence the assertion of the theorem follows.

2.3. The following theorem gives a characterization of convergent and Cauchy sequences with respect to any of the equivalent norms (2.5) which is independent of the special decomposition $\Pi_\varkappa = P \oplus N$.

Theorem 2.4. (i) The sequence (x_n), $x_n \in \Pi_\varkappa$, $n = 1,2,\ldots$, converges to $x_o \in \Pi_\varkappa$ with respect to an arbitrary norm (2.5) if and only if

$$(x_n,y) \to (x_o,y), \quad n \to \infty, \tag{2.6}$$

for all y of some total subset L of $\Pi_\varkappa$ and

$$(x_n,x_n) \to (x_o,x_o), \quad n \to \infty. \tag{2.7}$$

(ii) The sequence (x_n), $x_n \in \Pi_\varkappa$, $n = 1,2,\ldots$, is a Cauchy sequence with respect to an arbitrary norm (2.5) if and only if

$$(x_n - x_m, x_n - x_m) \to 0, \quad n,m \to \infty, \tag{2.8}$$

and the (x_n,y), $n = 1,2,\ldots$, form a Cauchy sequence for all elements y of some total subset L of $\Pi_\varkappa$.

The proof of this theorem is based on the following proposition (see [35]).

Lemma 2.1. Every dense linear manifold $M \subset \Pi_\varkappa$ contains a $\varkappa$-dimensional positive subspace.

Proof. Consider a decomposition (1.1) with a $\varkappa$-dimensional positive subspace Π_+, and choose an orthonormal basis $e_1,\ldots,e_\varkappa$ in Π_+. For arbitrary $\varepsilon > 0$ there are $f_1,\ldots,f_\varkappa \in M$ such that $\|e_j - f_j\| < \varepsilon$,

$j = 1,\dots,\varkappa$. If $\varepsilon > 0$ is sufficiently small, the vectors $f_1,\dots,f_\varkappa$ are linearly independent. Moreover, as the scalar product is continuous with respect to the norm $\|x\|$ (see 2.1), for sufficiently small $\varepsilon > 0$ the form $\sum_{j,k=1}^{\varkappa} (f_j,f_k)\, \xi_j \bar{\xi}_k$ is positive definite. Hence the linear span of $f_1,\dots,f_\varkappa$ is a $\varkappa$-dimensional positive subspace of M.

<u>Proof of Theorem 2.4.</u> The necessity of the conditions in (i) and (ii) follows immediately from (2.4). Suppose the sequence (x_n) has the properties (2.6) and (2.7) for all $y \in L$ or, what is the same, for all $y \in M$ where M is the linear manifold spanned by L. By Lemma 2.1, M has a $\varkappa$-dimensional positive subspace P. Let $e_1,\dots,e_\varkappa$ be an orthonormal basis of P and $\Pi_\varkappa = P \oplus N$ a corresponding decomposition of $\Pi_\varkappa$ (see Lemma 1.1). Consider

$$x_n = x_n^+ + x_n^- = \xi_{n1} e_1 + \dots + \xi_{n\varkappa} e_\varkappa + x_n^- , \quad x_n^- \in N ,$$

$n = 0,1,\dots$. Condition (2.6) implies $\xi_{nj} \to \xi_{oj}$, $n \to \infty$, $j = 1,\dots,\varkappa$, and hence $|x_n^+ - x_o^+| \to 0$. But (2.7) can be written as

$$|x_n^+|^2 - |x_n^-|^2 \to |x_o^+|^2 - |x_o^-|^2 ,$$

and we also get

$$|x_n^-| \to |x_o^-| \tag{2.9}$$

and $|x_n| \to |x_o|$, $n \to \infty$. Now (2.6) extends easily to all $y \in \Pi_\varkappa$, and for arbitrary $y^- \in N$ it follows that

$$(x_n^-, y^-) \to (x_o^-, y^-) \quad \text{if} \quad n \to \infty .$$

Together with (2.9) this gives $|x_n^- - x_o^-| \to 0$ as $n \to \infty$, and the statement follows.

If the sequence (x_n) satisfies the conditions in (ii), with the same notation as in the proof of (i) it follows that for $j = 1,\dots,\varkappa$ the same is true for (x_n^+). Now (2.8) implies the same for (x_n^-), and the statement follows.

<u>2.4.</u> It is well known that any incomplete unitary space H has a completion; to be more precise, that it can be imbedded isometrically in a complete unitary space H_1 the elements of which are the fundamental sequences of the space H. In this connection equivalent fundamental sequences are regarded as identical, i.e., they define the same element of H_1. If $X = (x_n)$ and $Y = (y_n)$ are two fundamental sequences of

vectors of the space H, then, regarding X and Y as elements of H_1, we define their scalar product by

$$(X,Y) = \lim_{n\to\infty} (x_n, y_n) .$$

The existence of this limit is easily verified, and it is as easily seen that the definition of (X,Y) is invariant under a change of the sequences (x_n) and (y_n) respectively within the equivalence classes, and that it has all the usual properties of a scalar product.

What is particularly important for us is the fact that the condition $(x,x) > 0$ for all $x \in H$ $(x \neq \theta)$ implies that $(X,X) > 0$ for all $X \in H_1$ $(X \neq \Theta)$, where Θ is the zero of the space H_1. In fact, if

$$0 = (X,X) = \lim_{n\to\infty} (x_n, x_n) = \lim_{n\to\infty} \|x_n\|^2 ,$$

then $x_n \to \theta$ as $n \to \infty$, and the sequence (x_n) defines the element $X = \Theta$.

Clearly the analogous proposition holds for the case that $(x,x) < 0$ if $x \neq \theta$.

These remarks enable us to establish without trouble the corresponding theorem on the completion of a space with an indefinite metric.

Theorem 2.5. A space E, satisfying axioms I - V of § 1 can always be completed in such a way that it becomes a space of the type $\Pi_\varkappa$ satisfying axioms I - VI.

Proof. In fact, if E is represented in the form $E = P \oplus N$, where P is a positive $\varkappa$-dimensional subspace of E (see Lemma 1.1), then we can complete the negative subspace N so that it becomes a complete unitary space N_1. As a result of this completion the scalar square of any vector different from zero in N_1 is strictly negative. Hence, in the space $\Pi_\varkappa = P \oplus N_1$ the axioms I - V will remain valid, and, in addition, it will be complete, as was required by axiom VI.

Let E again be a linear space satisfying only the axioms I - V of § 1. With a decomposition $E = P \oplus N$ a norm (2.3) can be introduced in E. Theorem 2.4 gives the following characterization of convergent and of Cauchy sequences with respect to this norm:

(i) The sequence (x_n), $x_n \in E$, converges to $x_0 \in E$ if and only if for $n \to \infty$ we have $(x_n, x_n) \to (x_0, x_0)$ and $(x_n, y) \to (x_0, y)$ for all $y \in E$.

(ii) The sequence $(x_n), x_n \in E$, is a Cauchy sequence if and only if $(x_n - x_m, x_n - x_m) \to 0$, $n,m \to \infty$, and the (x_n,y), $n = 1,2,\ldots$, form a Cauchy sequence for all $y \in E$.

Now the completion $\Pi_\varkappa$ of E (see Theorem 2.5) can also be introduced as the set of all sequences (x_n) of elements of E with the properties in (ii).

§ 3. Linear manifolds with degenerate and nondegenerate scalar products

3.1. In what follows we shall always suppose that together with the scalar product (x,y) $(x,y \in \Pi_\varkappa)$ in $\Pi_\varkappa$ there is given one of the mutually equivalent norms (see Theorem 2.3) resulting from a decomposition $\Pi_\varkappa = \Pi_+ \oplus \Pi_-$. This norm we write $\|x\|$ $(x \in \Pi_\varkappa)$ (cf. (2.3)). All concepts like limits, Cauchy sequences, closure, continuity of functionals and operators in $\Pi_\varkappa$ must be understood with respect to the norm $\|x\|$.

With the fixed decomposition $\Pi_\varkappa = \Pi_+ \oplus \Pi_-$ (and the corresponding H-scalar product $[x,y]$; see (2.1)) we define a linear operator J in $\Pi_\varkappa$: If $x = x_+ + x_-$ $(x_+ \in \Pi_+, x_- \in \Pi_-)$, then $Jx = x_+ - x_-$. The following properties of J are immediate consequences of this definition:

1°. $(Jx,y) = (x,Jy) = [x,y]$;

2°. $[Jx,y] = [x,Jy] = (x,y)$;

3°. $J(Jx) = J^2x = x$;

4°. $\|Jx\| = \|x\|$

whenever $x,y \in \Pi_\varkappa$.

We shall call the orthogonal complement of a set $M \subset \Pi_\varkappa$ with respect to the H-scalar product the <u>H-orthogonal complement</u> of M, and denote it by $M^{[\perp]}$. It is obvious from this definition, Defintion 1.4 and 1° that we have

$$(JM)^{[\perp]} = M^{\perp}, \quad (JM)^{\perp} = M^{[\perp]} . \tag{3.1}$$

Moreover, the relation

$$JM^{[\perp]} = (JM)^{[\perp]} \tag{3.2}$$

holds. Indeed, for $x \in M^{[\perp]}$ we have from 2° and 3°

$$\{0\} = [M,x] = [JM,Jx] ; \tag{3.3}$$

therefore $Jx \in (JM)^{[\perp]}$ and $JM^{[\perp]} \subset (JM)^{[\perp]}$. On the other hand, the same relation implies that for $y = Jx \in (JM)^{[\perp]}$ we have $Jy \in M^{[\perp]}$ or $y \in JM^{[\perp]}$.

As a consequence of (3.1) and (3.2) we get $M^{\perp} = JM^{[\perp]}$; hence the dimensions of $M^{\perp}$ and $M^{[\perp]}$ are the same, even if M is a subspace and $M \cap M^{\perp} = \{\theta\}$. The word subspace, in agreement with the usual terminology, always denotes a closed linear manifold in $\Pi_{\varkappa}$.

The following theorem is an easy consequence of (3.1) and (3.2).

<u>Theorem 3.1.</u> If M is an arbitrary nonempty subset of $\Pi_{\varkappa}$, then $M^{\perp}$ is a subspace of $\Pi_{\varkappa}$ and $(M^{\perp})^{\perp}$ coincides with the closed linear span of M.

<u>Proof.</u> Using the first relation of (3.1), (3.2) and the second relation in (3.1), we get

$$(M^{\perp})^{\perp} = ((JM)^{[\perp]})^{\perp} = (JM^{[\perp]})^{\perp} = (M^{[\perp]})^{[\perp]} .$$

But it is well-known in Hilbert space theory that the right hand side is the closed linear span of M. The first relation in (3.1) shows that $M^{\perp}$ is a subspace.

<u>Corollary.</u> If M is a subspace of $\Pi_{\varkappa}$, then $(M^{\perp})^{\perp} = M$.

<u>3.2.</u> As L.S. Pontrjagin [35] observed, among all the subspaces of the space $\Pi_{\varkappa}$ a special role is played by those on which the scalar product (x,y) $(x,y \in \Pi_{\varkappa})$ does not degenerate.

<u>Definition 3.1.</u> We say that on the linear manifold L in $\Pi_{\varkappa}$ the scalar product (x,y) $(x,y \in L)$ <u>degenerates</u> if L contains at least one vector $x_0 \neq \theta$ orthogonal to all of L, i.e., $x_0 \perp L$. Such a vector x_0 we call an <u>isotropic vector</u> of the linear manifold L (cf. A.I. Mal'cev [31]).

The set of all isotropic vectors of a linear manifold L becomes a linear manifold M when the zero vector θ is adjoined to it. On M the scalar product is identically zero: $M \subset L$, $M \perp L$. From Lemma 1.2 it follows that the dimension of M is not greater than $\varkappa$. The linear manifold M is called the <u>isotropic subspace</u> of L.

If L is a subspace of $\Pi_{\varkappa}$, its isotropic subspace G coincides with the isotropic subspace of $L^{\perp}$. Indeed, we have $G = L \cap L^{\perp}$, and the Corollary of Theorem 3.1 implies that the right hand side equals $L^{\perp} \cap (L^{\perp})^{\perp}$, which is the isotropic subspace of $L^{\perp}$.

If L is a linear manifold on which the scalar product does not degenerate, then the relations $x \in L$ and $x \perp L$ always imply that $x = \theta$. This property is obeyed, for example, by the positive and negative linear manifolds of $\Pi_\varkappa$.

Further, the following simple remark follows from Lemma 1.2:

Any linear manifold $L \subset \Pi_\varkappa$ containing a $\varkappa$-dimensional positive subspace P is a linear manifold with nondegenerating scalar product.

3.3. Axiom III of § 1 implies that on the whole of $\Pi_\varkappa$ the scalar product does not degenerate. From this it follows that in any decomposition $\Pi_\varkappa = L \oplus M$ of $\Pi_\varkappa$ into a direct sum of orthogonal subspaces L and M the scalar product does not degenerate on either of the components.

The last statement has in a certain sense a converse, namely:

Theorem 3.2.[1]) If the scalar product does not degenerate on the subspace $L \subset \Pi_\varkappa$, then it also does not degenerate on its orthogonal complement M, and $\Pi_\varkappa$ is decomposed into the direct orthogonal sum $\Pi_\varkappa = L \oplus M$.

Proof. Using in part an idea of L.S. Pontrjagin, we split the proof into three steps.

a) Let L be a negative subspace of finite dimension p. Choose in L an orthonormal basis $\{h_j\}_{j=1}^{p}$, $(h_j, h_k) = -\delta_{jk}$ $(j,k = 1,\dots,p)$. Any element $x \in \Pi_\varkappa$ can then be put in the form

$$x = \sum_{j=1}^{p} (x, h_j) h_j + y ,$$

where, as is easily seen, $y \in M$. From this we get the decomposition $\Pi_\varkappa = L \oplus M$, and from this and the conditions of the theorem it follows that the scalar product does not degenerate on M.

b) Let L be an infinite-dimensional negative subspace. Consider the intersection $F = L \cap \Pi_-$. F, being the intersection of two subspaces, is a subspace of the unitary space Π_-. This implies that the scalar product does not degenerate on F. Consider two decompositions

[1]) This theorem corresponds to Theorem 2 of the article [35] of L.S. Pontrjagin.

$$\Pi_- = F \oplus \Pi'_- , \qquad L = F \oplus L' ,$$

where Π'_- and L' are the orthogonal complements of F in Π_- and L, respectively. The direct sum $\Pi'_\varkappa = \Pi_+ \oplus \Pi'_-$ obviously satisfies all the axioms I - VI of § 1, i.e., is a space of type $\Pi_\varkappa$. Moreover, $\Pi_\varkappa = F \oplus \Pi'_\varkappa$. L' is a negative subspace of the space $\Pi'_\varkappa$. We show as in the proof of Lemma 1.2 that L' is finite-dimensional. In fact, it is clear that $L' \cap \Pi_- = \{\theta\}$. Hence, if

$$x, y \in L', \quad x = x_+ + x_-, \quad y = y_+ + y_-$$

and $x_+ = y_+$, then $x - y = x_- - y_- = \theta$, i.e., $x = y$. In consequence, each element $x \in L'$ is completely determined by its component x_+ lying in the $\varkappa$-dimensional space Π_+. This implies that the dimension of L' does not exceed $\varkappa$. On the basis of the previous section of the proof we get that

$$\Pi'_\varkappa = L' \oplus M, \quad \Pi_\varkappa = F \oplus \Pi'_\varkappa = F \oplus L' \oplus M = L \oplus M ,$$

as required.

c) The case of an arbitrary subspace L can be deduced from the preceding without trouble. For, let P be a p-dimensional subspace of L with the largest dimension of any positive subspace in L. It is clear that $0 \leq p \leq \varkappa$. We have investigated the case $p = 0$. Let $p > 0$. The subspace L in this case satisfies axioms I - VI with p instead of $\varkappa$. By Lemma 1.1 we have $L = P \oplus L_1$, where L_1 is a negative subspace with nondegenerating scalar product. On the other hand, choosing in P an orthonormal basis and repeating the considerations in section a) of the present proof, we can show that $\Pi_\varkappa = P \oplus \Pi_{\varkappa-p}$, where $\Pi_{\varkappa-p}$ satisfies axioms I - VI with $\varkappa - p$ instead of $\varkappa$. It is also clear that $L_1 \subset \Pi_{\varkappa-p}$, and on the basis of step b), $\Pi_{\varkappa-p} = L_1 \oplus M$, i.e., $\Pi_\varkappa = P \oplus L_1 \oplus M = L \oplus M$.

The theorem is completely proved.

Corollary 1. If the scalar product does not degenerate on the subspace $L \subset \Pi_\varkappa$, then for any decomposition $L = P \oplus N$ into positive and negative components (cf. Lemma 1.1) we can find a decomposition $\Pi_\varkappa = \Pi_+ \oplus \Pi_-$ so that $P \subset \Pi_+$ and $N \subset \Pi_-$.

To prove this it is sufficient to represent $\Pi_\varkappa$ in the form $\Pi_\varkappa = L \oplus M$, and to decompose M by Lemma 1.1 into positive and negative components: $M = P_1 \oplus N_1$. If now we put $\Pi_+ = P \oplus P_1$ and $\Pi_- = N \oplus N_1$, then $\Pi_\varkappa = \Pi_+ \oplus \Pi_-$ as required.

Corollary 2. If the scalar product does not degenerate on the subspace $L \subset \Pi_\varkappa$, then L with the scalar product (x,y), $x,y \in L$, is a space of type $\Pi_{\varkappa'}$ for some $\varkappa' \leq \varkappa$, that is, it satisfies axioms I - VI of § 1 for $\varkappa'$ instead of $\varkappa$.

3.4. In the following we shall have to deal with linear manifolds L in $\Pi_\varkappa$ on which the scalar product degenerates; that is, the isotropic subspace $G = L \cap L^\perp$ of L does not consist only of the vector θ.

Theorem 3.3. If on the linear manifold L the scalar product degenerates, and G is the isotropic subspace of L, then L may be decomposed into the direct sum $L = G \oplus L_1$, where L_1 is a linear manifold with nondegenerating scalar product.

Proof. We consider L and G as linear manifolds of a Hilbert space H with the H-scalar product $[x,y]$ $(x,y \in \Pi_\varkappa)$ (cf. (2.1)). As G is a finite-dimensional subspace, we obtain without difficulty the decomposition $L = G[\dot{+}]L_1$, where the symbol $[\dot{+}]$ denotes the H-orthogonal direct sum.

As $G \perp L_1$, however, we have that $L = G \oplus L_1$. It remains to show that the scalar product does not degenerate on L_1. If L_1 contained an isotropic vector x_0, then from the relation $x_0 \perp L_1$ and $x_0 \in L_1 \perp G$ it would follow that x_0 is an isotropic vector for all L. At the same time, $x_0 \notin G$, and this contradicts the definition of the isotropic subspace of a linear manifold L (see 3.2).

Note. The decomposition $L = G \oplus L_1$ obtained in the proof of Theorem 3.3 is, generally speaking, only one of the possible decompositions of this form. It is clear that each of the decompositions $\Pi_\varkappa = \Pi_+ \oplus \Pi_-$ generates its H-scalar product, and this in its turn gives rise to a decomposition of the linear manifold L of the form considered above.

Even if the scalar product does not degenerate on the linear manifold $L \subset \Pi_\varkappa$, which is not a subspace (i.e. L is not closed), it can degenerate on $\bar{L}$ (the closure of L). This fact will be shown incidentally below (see § 6, Example 6.1).

3.5. In this section we prove a generalization of Theorem 3.2 for an arbitrary subspace $L \subset \Pi_\varkappa$.

Lemma 3.1. If G is a neutral subspace of $\Pi_\varkappa$, then there exists a subspace F $(\subset \Pi_\varkappa)$ of the same dimension as G which has the following properties:

a) F is neutral;

b) no vector of G different from θ is orthogonal to F;

c) no vector of F different from θ is orthogonal to G;

d) the scalar product does not degenerate on the direct sum $G \dot{+} F$.

Proof. With the operator J defined in 3.1 we can take the subspace $F = JG$. Indeed, the properties 1° and 3° of J imply $[Jx,Jy] = [x,y] = 0$ $(x,y \in G)$; that is, a). Further, if $(\theta \neq)$ $g \in G$, then $f = Jg \in F$ and $(g,f) = (g,Jg) = [g,g] > 0$, so that g is not orthogonal to f, which proves b). Moreover, since $J^2 = I$, we have $G = JF$, and this implies c) by a completely analogous proof. Finally, assertion d) follows easily from the preceding ones.

Note 1. The subspace F with the properties a) - d) is not uniquely determined (cf. the note after Theorem 3.3).

Note 2. Since G and F are finite-dimensional spaces of equal dimension m $(\leq \varkappa)$, for any basis $g_1,\dots,g_m \in G$ there exists a basis $f_1,\dots,f_m \in F$ such that $(g_j,f_k) = \delta_{jk}$ $(j,k = 1,\dots,m)$.

Definition 3.2. If G is a neutral subspace of $\Pi_\varkappa$ and the subspace F has the properties a) - d) of Lemma 3.1, then G and F are said to be skewly linked.

Now the following generalization of Theorem 3.2 can be proved ([19b], [9]).

Theorem 3.4. Let L be an arbitrary subspace of $\Pi_\varkappa$ and denote by M the orthogonal complement $L^\perp$ and by G the isotropic part $L \cap L^\perp$ of L.

(i) If we choose decompositions $L = L_1 \oplus G$ and $M = M_1 \oplus G$ (see Theorem 3.3), then there exists at least one subspace $F \subset \Pi_\varkappa$ which is skewly linked with G and such that

$$\Pi_\varkappa = L_1 \oplus (G \dot{+} F) \oplus M_1 \,. \tag{3.4}$$

(ii) If F is an arbitrary subspace of $\Pi_\varkappa$ which is skewly linked with G then there exist unique decompositions $L = L_1 \oplus G$ and $M = M_1 \oplus G$ such that (3.4) holds.

Proof. (i) On $L_1 \oplus M_1$ the scalar product does not degenerate. By Theorem 3.2 the same is true for its orthogonal complement E, and we have $\Pi_\varkappa = L_1 \oplus M_1 \oplus E$. From Corollary 2 of Theorem 3.2, E is a space of type $\Pi_{\varkappa'}$, $0 \leqq \varkappa' \leqq \varkappa$, and G is evidently a neutral subspace of E. Then there exists a subspace $F \subset E$ which is skewly linked with G (Lemma 3.1), and we have

$$\Pi_\varkappa \supset L_1 \oplus M_1 \oplus (G \dotplus F) .$$

If x belongs to the orthogonal complement of the right-hand side, it follows that $x \perp M$, $x \perp L$ and $x \perp F$; that is, $x \in G$ and $x \perp F$, so that $x = \theta$. An application of the Corollary to Theorem 3.1 gives (3.4).

(ii) The scalar product does not degenerate on $G \dotplus F$. Denoting by H its orthogonal complement, we have $\Pi_\varkappa = (G \dotplus F) \oplus H$. An arbitrary $x \in L$ can be written as $x = g + f + h$, $g \in G$, $f \in F$, $h \in H$, and $(x,G) = \{0\}$ implies $f = \theta$; that is, $L \subset G \oplus H$. From $G \subset L$ we get $L = G \oplus L_1$ with some subspace $L_1 \subset H$. Evidently, the scalar product does not degenerate on L_1. In the same way, $M = G \oplus M_1$, where $M_1 \subset H$, the scalar product does not degenerate on M_1 and M_1 is orthogonal to L_1. Hence $\Pi_\varkappa \supset (G \dotplus F) \oplus L_1 \oplus M_1$, and the relation (3.4) follows in the same way as in (i).

Corollary 1. If G is a neutral subspace of $\Pi_\varkappa$, F is a subspace of $\Pi_\varkappa$ which is skewly linked with G and M denotes the orthogonal complement of G: $M = G^\perp$, then $\Pi_\varkappa = M \dotplus F$.

§ 4. Coordinate description of a separable space $\Pi_\varkappa$

4.1. Careful consideration of axiom VI of § 1 and the theorems of § 2 connected with it show that if in axiom VI we demand, in addition to the completeness, the separability of Π_-, then $\Pi_\varkappa$ is a separable space with respect to any of the equivalent norms (2.5).

In the case when $\Pi_\varkappa$ is a separable space it is sometimes convenient to describe its element by infinite coordinate vectors. To do this we shall begin with some fixed decomposition $\Pi_\varkappa = \Pi_+ \oplus \Pi_-$ and introduce orthogonal coordinates into Π_+ and Π_-. For this we choose in Π_+ an orthonormal basis $(e_j)_{j=1}^{\varkappa}$ and in Π_- an orthonormal basis $(e_j)_{j=\varkappa+1}^{\infty}$, that is,

$$(e_j,e_k) = 0 \quad (j \neq k); \quad (e_j,e_j) = \begin{cases} 1 & (j \leqq \varkappa) \\ -1 & (j > \varkappa) \end{cases} \quad (j,k = 1,2,\ldots).$$

The set $(e_j)_{j=1}^{\infty}$ is a basis for the space $\Pi_\varkappa$, i.e., any vector $x \in \Pi_\varkappa$ has a unique representation in the form of a convergent series

$$x = \xi_1 e_1 + \xi_2 e_2 + \dots , \tag{4.1}$$

or, for short,

$$x = (\xi_1, \xi_2, \dots), \text{ where } [x,x] = \sum_{j=1}^{\infty} |\xi_j|^2 < \infty . \tag{4.2}$$

On the other hand, any series (4.1) for which (4.2) holds is convergent, i.e., defines a vector $x \in \Pi_\varkappa$. If

$$x = (\xi_1, \xi_2, \dots), \quad \sum_{j=1}^{\infty} |\xi_j|^2 < \infty ,$$

$$y = (\eta_1, \eta_2, \dots), \quad \sum_{j=1}^{\infty} |\eta_j|^2 < \infty ,$$

then

$$(x,y) = \xi_1\bar{\eta}_1 + \dots + \xi_\varkappa\bar{\eta}_\varkappa - \xi_{\varkappa+1}\bar{\eta}_{\varkappa+1} - \xi_{\varkappa+2}\bar{\eta}_{\varkappa+2} - \dots , \tag{4.3}$$

$$[x,y] = \xi_1\bar{\eta}_1 + \dots + \xi_\varkappa\bar{\eta}_\varkappa + \xi_{\varkappa+1}\bar{\eta}_{\varkappa+1} + \xi_{\varkappa+2}\bar{\eta}_{\varkappa+2} + \dots , \tag{4.4}$$

$$\|x\| = \Big(\sum_{j=1}^{\infty} |\xi_j|^2\Big)^{\frac{1}{2}} . \tag{4.5}$$

Thus $\Pi_\varkappa$, considered as a space with the scalar product (4.4), is isomorphic (as is every separable Hilbert space) with the space ℓ_2. Considered as a Banach space with the norm (4.5) and the scalar product (4.3), $\Pi_\varkappa$ coincides with the space $H_\varkappa$ with indefinite metric introduced by L.S. Pontrjagin [35]. There is one inessential difference between $\Pi_\varkappa$ and $H_\varkappa$, namely that in Pontrjagins work the scalar product differs in sign from the scalar product (4.3).

4.2. The procedure which has been explained for introducing coordinates in $\Pi_\varkappa$ admits the following generalization, which is important in applications of the theory to concrete spaces.

Let an infinite Hermitian matrix

$$\|c_{jk}\|_{j,k=1}^{\infty} \qquad (c_{kj} = \overline{c_{jk}}, \quad j,k = 1,2,\dots)$$

be given, for which it is known that the form

$$\sum_{j,k=1}^{n} c_{jk}\, \xi_j \bar{\xi}_k \tag{4.6}$$

for all sufficiently large natural n contains just $\varkappa$ $(0 < \varkappa < n)$ positive squares. Consider now the linear space $\widetilde{\Pi}$ of sequences $x = (\xi_j)$, for each of which only a finite number of coordinates ξ_j (depending on x) are different from zero.

If $x, y \in \tilde{\Pi}$, and $x = (\xi_j)$, $y = (\eta_j)$ and $\xi_j = \eta_j = 0$ $(j > n)$ then the scalar product in $\tilde{\Pi}$ will be defined by

$$(x,y) = \sum_{j,k=1}^{n} c_{jk}\, \xi_j \bar{\xi}_k \,.$$

With this definition the axioms I and II are fulfilled. The presence of $\varkappa$ positive squares in (4.6) ensures that axiom IV is fulfilled, and the absence (for any n) of a larger number of positive quares is equivalent to axiom V.

Axiom III will be fulfilled if we require in addition that arbitrarily large values of n can be found for which the forms (4.6) are non-degenerate, i.e.,

$$\Delta_n = \det\| c_{jk}\|_{j,k=1}^{n} \neq 0 \,.$$

It remains only to complete $\tilde{\Pi}$ according to Theorem 2.5 in order to obtain a space $\Pi_\varkappa$.

We mention that for an arbitrary Hermitian matrix $\| c_{jk}\|_{j,k=1}^{\infty}$ the indefinite metric given by this matrix in the space Π may be non-degenerate, even if there is no infinite subsequence of minors Δ_n different from zero. This can be seen from the example

$$\| c_{jk}\|_{j,k=1}^{\infty} = \begin{bmatrix} 0 & 0 & 1 & 0 & 0 & 0 & 0 & 0 & 0 & \cdots \\ 0 & 0 & 0 & 0 & 1 & 0 & 0 & 0 & 0 & \cdots \\ 1 & 0 & 0 & 0 & 0 & 0 & 0 & 0 & 0 & \cdots \\ 0 & 0 & 0 & 0 & 0 & 0 & 1 & 0 & 0 & \cdots \\ 0 & 1 & 0 & 0 & 0 & 0 & 0 & 0 & 0 & \cdots \\ 0 & 0 & 0 & 0 & 0 & 0 & 0 & 0 & 1 & \cdots \\ 0 & 0 & 0 & 1 & 0 & 0 & 0 & 0 & 0 & \cdots \\ 0 & 0 & 0 & 0 & 0 & 0 & 0 & 0 & 0 & \cdots \\ 0 & 0 & 0 & 0 & 0 & 1 & 0 & 0 & 0 & \cdots \\ \vdots & \vdots & \vdots & \vdots & \vdots & \vdots & \vdots & \vdots & \vdots & \end{bmatrix}$$

for which $\Delta_n = 0$ $(n = 1,2,\ldots)$, but the corresponding metric is non-degenerate. However, if the number of positive squares of the form (4.6) is finite, this is impossible [18e].

Suppose the infinite matrix $\| c_{jk}\|_{j,k=1}^{\infty}$ has finite rank r (that is, r is the maximal order of the nonzero minors of this matrix), and the form (4.6) for all sufficiently large n has $\varkappa$ positive squares $(r \geqq \varkappa)$. Then the metric given in $\tilde{\Pi}$ by this matrix degenerates. Denote the isotropic part of $\tilde{\Pi}$ by $\tilde{\Pi}^{(o)}$. Then it is easy to check that the linear manifold $\tilde{\Pi}/\tilde{\Pi}^{(o)}$ is an r-dimensional space of type $\Pi_{\varkappa}$.

This can also be seen immediately from the fact, that for the Hermitian matrix $\| c_{jk} \|_{j,k=1}^{\infty}$ of finite rank r $(\geqq \varkappa)$ there is a nonzero main minor of order r. After changing the enumeration of rows and columns we can suppose that $\Delta_r \neq 0$. Then, evidently,

$$\begin{vmatrix} c_{11} & c_{12} & \cdots & c_{1r} & c_{1k} \\ c_{21} & c_{22} & \cdots & c_{2r} & c_{2k} \\ \cdots & \cdots & \cdots & \cdots & \cdots \\ c_{r1} & c_{r2} & \cdots & c_{rr} & c_{rk} \\ c_{j1} & c_{j2} & \cdots & c_{jr} & c_{jk} \end{vmatrix} = 0 \quad (j,k = r+1, r+2, \ldots) ,$$

and because of a well-known formula of Sylvester (see [15]) we have

$$\sum_{j,k=1}^{n} c_{jk} \xi_j \bar{\xi}_k = -\frac{1}{\Delta_r} \begin{vmatrix} c_{11} & c_{12} & \cdots & c_{1r} & \bar{X}_1 \\ c_{21} & c_{22} & \cdots & c_{2r} & \bar{X}_2 \\ \cdots & \cdots & \cdots & \cdots & \cdots \\ c_{r1} & c_{r2} & \cdots & c_{rr} & \bar{X}_r \\ X_1 & X_2 & \cdots & X_r & 0 \end{vmatrix}$$

where

$$X_j = \sum_{k=1}^{n} c_{jk} \xi_k \quad (j = 1,2,\ldots,r;\ n > r) . \tag{4.7}$$

Thus, in the space $\tilde{\Pi}$ defined above the metric is in this case given essentially by the form

$$-\frac{1}{\Delta_r} \begin{vmatrix} c_{11} & c_{12} & \cdots & c_{1r} & \bar{X}_1 \\ c_{21} & c_{22} & \cdots & c_{2r} & \bar{X}_2 \\ \cdots & \cdots & \cdots & \cdots & \cdots \\ c_{r1} & c_{r2} & \cdots & c_{rr} & \bar{X}_r \\ X_1 & X_2 & \cdots & X_r & 0 \end{vmatrix} = \sum_{j,k=1}^{r} d_{jk} X_j \bar{X}_k . \tag{4.8}$$

Here we identify all the vectors (ξ_j) which correspond by (4.7) to the same r-dimensional vector $(X_1, \ldots, X_r)$.

The r-dimensional space $\tilde{\Pi} = \Pi_{\varkappa}$ obtained in this way evidently needs no completion. If we choose the basis in this space in the same way as in 4.1 for the infinite-dimensional space $\Pi_{\varkappa}$, then with this basis the

Hermitian form (4.8) is brought into canonical form, i.e., the scalar product

$$\sum_{j,k=1}^{r} d_{jk} X_j \bar{X}_k$$

is written in the form (4.3) with a finite number (r) of components on the right-hand side.

C H A P T E R   2

I S O M E T R I C   A N D   H E R M I T I A N   O P E R A T O R S   I N   A   S P A C E $\Pi_\varkappa$

§ 5. Linear functionals in $\Pi_\varkappa$. The adjoint operator

5.1. Since $\Pi_\varkappa$ is a Hilbert space, additionally equipped with an indefinite scalar product, the concepts of linear functional and linear operator in $\Pi_\varkappa$ do not need special definition. However, on making a comparison with the usual Hilbert space, all the properties of linear functionals and operators which are connected with the concept of scalar product have to be reconsidered, as was done in part by Pontrjagin [35].

First of all, we introduce those facts and definitions which can be carried over without trouble from the theory of operators in Hilbert space.

Lemma 5.1. If $F(x)$ is a linear functional in $\Pi_\varkappa$, then there is a unique vector $f \in \Pi_\varkappa$ such that $F(x) = (x,f)$ for all $x \in \Pi_\varkappa$.

Proof. We start from a decomposition (1.1), the corresponding H-scalar product and the operator J (see 3.1). By a well-known theorem of F. Riesz there is a vector $g \in \Pi_\varkappa$ such that $F(x) = [x,g]$, $x \in \Pi_\varkappa$. Taking $f = Jg$, we get $F(x) = (x,f)$, $x \in \Pi_\varkappa$. The uniqueness of the vector f follows from the fact that the scalar product does not degenerate on $\Pi_\varkappa$.

Lemma 5.1 shows that the weak convergence of a sequence in $\Pi_\varkappa$, defined in 1.5, is consistent with the usual definition in a Banach space. In the following the weak convergence of the sequence (x_n), $x_n \in \Pi_\varkappa$, $n = 1,2,\dots$, to an element $x_0 \in \Pi_\varkappa$ will be denoted by $x_n \xrightarrow{w} x_0$.

5.2. Let A be a linear operator in $\Pi_\varkappa$. The domain of A will be denoted by D(A) and the range by R(A). Symbolically we can write AD(A) = R(A). D(A) and R(A) are linear manifolds.

If A is a linear operator with domain D(A) dense in $\Pi_\varkappa$, and D_1 is the set of all vectors $y \in \Pi_\varkappa$ for each of which there is a corresponding vector $z \in \Pi_\varkappa$ such that for all $x \in D(A)$ $(Ax,y) = (x,z)$, then D_1 is clearly a linear manifold, and since $\overline{D(A)} = \Pi_\varkappa$, the vector z defined by the vector $y \in D_1$ is uniquely determined. In consequence, putting $z = A^*y$, we get that A^* is a linear operator with domain $D(A^*) = D_1$, where, in general $D(A^*) \neq \Pi_\varkappa$. The operator A^* is called adjoint to the operator A.

From the definition it is easy to see that the usual properties hold:

$$(A + B)^* \supset A^* + B^*, \quad (\lambda A)^* = \bar{\lambda} A^* . \tag{5.1}$$

Choosing a fixed H-scalar product $[x,y]$, $x,y \in \Pi_\varkappa$, and the corresponding operator J (see 3.1), we denote the adjoint of A with respect to this H-scalar product by $A^{[*]}$. Then it follows easily from the definitions that

$$A^* = JA^{[*]} J, \quad A^{[*]} = J A^* J . \tag{5.2}$$

The relations (5.1) are also immediate consequences of (5.2) and the corresponding properties of adjoint operators in Hilbert space. Also the following proposition can be proved directly or follows immediately from (5.2) and the corresponding Hilbert space statement.

The adjoint operator A^* is closed, that is, $f_n \in D(A^*)$, $n = 1,2,\dots$, $f_n \to f$ and $A^* f_n \to g$ imply $f \in D(A^*)$ and $A^*f = g$.

Actually, the (formally) stronger proposition that A^* is weakly closed follows; this means that $f_n \in D(A^*)$, $n = 1,2,\dots$, $f_n \xrightarrow{w} f$ and $A^* f_n \xrightarrow{w} g$ $(n \to \infty)$ imply $f \in D(A^*)$ and $A^*f = g$. Since convergence in norm of vectors implies their weak convergence, weak closedness of an operator implies that it is closed.

5.3. If the operator A is continuous (bounded) and $D(A) = \Pi_\varkappa$, then the adjoint operator is also defined on $\Pi_\varkappa$: $D(A^*) = \Pi_\varkappa$. This follows from (5.2) and the fact that $D(A^{[*]}) = \Pi_\varkappa$. Moreover, A^* is also continuous, and for every densely defined operator B the first relation in (5.1) becomes

$$(A + B)^* = A^* + B^* . \tag{5.1'}$$

For continuous operators A and B defined on $\Pi_\varkappa$ it is also clear that $(A^*)^* = A$ and $(AB)^* = B^*A^*$.

Now let A^{-1} denote the inverse operator of A, both not necessarily bounded. Then, if $\overline{D(A)} = \overline{D(A^{-1})} = \Pi_\varkappa$, in the same way as in the

Hilbert space case[1]) or from (5.2) it follows that $(A^*)^{-1}$ exists and that

$$(A^*)^{-1} = (A^{-1})^* . \tag{5.3}$$

In particular, this holds for bounded operators A and A^{-1} (defined on all $\Pi_\varkappa$).

§ 6. Some classes of linear operators in $\Pi_\varkappa$

6.1. Definition 6.1. The operator A, which has as domain a linear manifold D(A) dense in $\Pi_\varkappa$, is called Hermitian[2]) if for any $x,y \in D(A)$

$$(Ax,y) = (x,Ay) . \tag{6.1}$$

Note that (6.1) implies the linearity of A. Indeed, for any complex λ, μ and any $x,y,z \in D(A)$ we have

$$(A(\lambda x + \mu y),z) = (\lambda x + \mu y, Az) = \lambda(x,Az) + \mu(y,Az)$$
$$= \lambda(Ax,z) + \mu(Ay,z) = (\lambda Ax + \mu Ay, z) ,$$

whence, as D(A) is dense in $\Pi_\varkappa$ and the scalar product does not degenerate, it follows that

$$A(\lambda x + \mu y) = \lambda Ax + \mu Ay .$$

As for Hermitian operators in Hilbert space, in order that a linear operator A ($\overline{D(A)} = \Pi_\varkappa$) be Hermitian it is necessary and sufficient that for all $x \in D(A)$ the form (Ax,x) be real.

Only the sufficiency of the condition needs to be verified, and this follows immediately from the identity

$$(Ax,y) = \frac{1}{4}\{(A[x + y],x + y) - (A[x - y],x - y)\} +$$
$$+ \frac{i}{4}\{(A[x + iy],x + iy) - (A[x - iy],x - iy)\} \quad (x,y \in D(A)).$$

If A is a Hermitian operator, then, as is easily seen, A^* is an extension of A; that is, $D(A) \subset D(A^*)$; $Ax = A^*x$ $(x \in D(A))$. From this

[1]) Cf., e.g., [2].

[2]) Here we depart from the terminology of L.S. Pontrjagin, which one of the authors retained in the notes [18a,c].
An operator of the type introduced by our Definition 6.1 is called symmetric by Pontrjagin, and operators which are selfadjoint in the sense of our Definition 6.2 are called Hermitian by Pontrjagin (following A.I. Plesner).

and from the fact that A^* is closed, as usual, it follows that each Hermitian operator has a minimal closed Hermitian extension, i.e., it can be closed.

Definition 6.2. The linear operator A with domain D(A) $(\overline{D(A)} = \Pi_\varkappa)$ is called selfadjoint if $A^* = A$, i.e., $D(A) = D(A^*)$ and $Ax = A^*x$ $(x \in D(A))$.

From this definition it follows that a selfadjoint operator is Hermitian and closed.

If in Definition 6.2 we required instead of linearity that A should be a Hermitian operator, then the linearity of A would again follow from this.

An example of a selfadjoint operator in $\Pi_\varkappa$ is the operator J introduced in 3.1. With this operator J and the corresponding H-scalar product $[x,y]$, $x,y \in \Pi_\varkappa$, the definition of a Hermitian (selfadjoint) operator is equivalent to the requirement that JA should be H-Hermitian (H-selfadjoint); that is Hermitian (selfadjoint) in the H-scalar product.[1])

6.2. **Let A again be an Hermitian operator in $\Pi_\varkappa$. Then, by definition,** $\overline{D(A)} = \Pi_\varkappa$. By Lemma 2.1 the $\varkappa$-dimensional positive subspace Π_+ of the decomposition (1.1) can be chosen in D(A): $\Pi_+ \subset D(A)$. Then the restriction of A to Π_+ is a finite-dimensional, hence continuous linear operator, which we shall write as follows:

$$Ax_+ = (Ax_+)_+ + (Ax_+)_- = A_{11}\, x_+ + A_{21}\, x_+ \; .$$

Evidently, A_{11} (A_{21}) is a continuous linear operator from Π_+ into Π_+ (from Π_+ into Π_-, respectively). An element $x \in \Pi_\varkappa$ belongs to D(A) if and only if in its decomposition $x = x_+ + x_-$ according to (1.1) we have $x_- \in D(A)$. Therefore $JD(A) = D(A)$ and $D(A) \cap \Pi_-$ is dense in Π_-. If $x_- \in D(A) \cap \Pi_-$, then the decomposition

$$Ax_- = (Ax_-)_+ + (Ax_-)_- = A_{12}\, x_- + A_{22}\, x_-$$

defines two linear operators A_{12} and A_{22} from $D(A) \cap \Pi_-$ into Π_+ and Π_-, respectively. Moreover, for arbitrary $x_- \in D(A) \cap \Pi_-$ and $y_+ \in \Pi_+$ we have

[1])This is the origin of a terminology different from ours, in which unitary and selfadjoint operators in $\Pi_\varkappa$, regarded as operators in a Hilbert space, are called J-unitary and J-selfadjoint, respectively (cf. also J-increasing [21a] and J-nonincreasing [36] matrices, comp. with our Definition 6.6).

$$(A_{12}x_-,y_+) = (Ax_-,y_+) = (x_-,Ay_+) = (x_-,A_{21}y_+) .$$

With the H-scalar product (2.1) on Π_+ and Π_- this can be written as

$$[A_{12}x_-,y_+] = -[x_-,A_{21}y_+] \quad (x_- \in D(A) \cap \Pi_-,\ y_+ \in \Pi_+) . \tag{6.2}$$

Denoting the adjoint of A_{21} with respect to the H-scalar product on Π_+ and Π_- again by $A_{21}^{[*]}$, (6.2) shows that A_{12} is the restriction of $-A_{21}^{[*]}$ to $D(A) \cap \Pi_-$, hence continuous.

In the following it is sometimes convenient to write, instead of $x = x_+ + x_-$,

$$x = \begin{bmatrix} x_+ \\ x_- \end{bmatrix} .$$

Then the Hermitian operator A has the matrix representation

$$A = \begin{bmatrix} A_{11} & -A_{21}^{[*]} \\ A_{21} & A_{22} \end{bmatrix} . \tag{6.3}$$

It is easy to see that A_{11} is an H-selfadjoint operator in Π_+: $A_{11} = A_{11}^{[*]}$, and A_{22} is an H-Hermitian operator in Π_-: $A_{22} \subset A_{22}^{[*]}$. Evidently, A is closed (selfadjoint) if and only if A_{22} is closed (H-selfadjoint, respectively).

From (6.3) we have the following decomposition of A:

$$A = A_o + B, \quad A_o = \begin{bmatrix} A_{11} & 0 \\ 0 & A_{22} \end{bmatrix} , \quad B = \begin{bmatrix} 0 & -A_{21}^{[*]} \\ A_{21} & 0 \end{bmatrix} \tag{6.4}$$

Here A_o is H-Hermitian in $\Pi_\varkappa$ and B is a finite-dimensional operator: $\dim R(B) \leq 2\varkappa$.

We remind the reader that the <u>spectrum</u> $\sigma(T)$ of a linear operator T with domain D(T) is the complement of the set $\rho(T)$ of all its regular points: A <u>regular point</u> of T is a complex number λ with the property that the operator $T - \lambda I$ (I is the identity or "unit" operator) maps D(T) onto the whole of $\Pi_\varkappa$ and has a bounded inverse operator $(T - \lambda I)^{-1}$. The spectrum of an operator T contains all its <u>eigenvalues</u>, which form the so-called <u>point spectrum</u> $\sigma_p(T)$.

As usual (see e.g. [13]), by $\sigma_c(T)$ and $\sigma_r(T)$ we shall denote the <u>continuous</u> and <u>residual spectrum</u>, respectively, of the operator T. We remind that $\sigma_c(T)$ is the set of all complex λ for which there exists a densely defined unbounded inverse operator $(T - \lambda I)^{-1}$, and $\sigma_r(T)$ is the set of all those λ for which $\lambda \notin \sigma_p(T)$ and $\overline{(T - \lambda I)D(T)} \neq \Pi_\varkappa$. The sets $\sigma_p(T)$, $\sigma_c(T)$ and $\sigma_r(T)$ are mutually disjoint, and

$$\sigma_p(T) \cup \sigma_c(T) \cup \sigma_r(T) = \sigma(T) .$$

If λ is an eigenvalue of T, the linear manifold S_λ (or $S_\lambda(T)$) of all vectors x for which there exists a positive integer r (dependent on x) such that $(T - \lambda I)^r x = \theta$ is called the <u>root manifold</u> of T corresponding to the eigenvalue λ.

In the following it will be convenient to have a special notation for some subset of the point spectrum $\sigma_p(T)$. The <u>discrete spectrum</u> $\sigma_d(T)$ is by definition the set of all eigenvalues λ of T with the following properties:

(i) λ is an isolated point of $\sigma(T)$;

(ii) the root manifold S_λ is finite-dimensional;

(iii) $\Pi_\varkappa$ decomposes as the direct sum of S_λ and a subspace S'_λ: $\Pi_\varkappa = S_\lambda \dotplus S'_\lambda$, such that $T - \lambda I$ maps $D(T) \cap S'_\lambda$ onto S'_λ.

In this definition the condition (i) also follows from (ii) and (iii) (see e.g. [17b]). In [17b] the points of $\sigma_d(T)$ are called <u>normal eigenvalues</u>.

A complex number λ is called a <u>point of regular type</u> of the closed operator T if λ is not an eigenvalue and $R(T - \lambda I)$ is closed. Then the inverse $(T - \lambda I)^{-1}$ exists and is continuous, but not necessarily everywhere defined. This definition of a point of regular type is equivalent to the following: There exists a positive number $k = k(\lambda)$, such that $\|Tx - \lambda x\| \geqq k\|x\|$ for all $x \in D(T)$.

Now let A again be an Hermitian operator in $\Pi_\varkappa$ ($\overline{D(A)} = \Pi_\varkappa$). If $Af = \lambda f$, $Ag = \mu g$, $\lambda \neq \overline{\mu}$, then from the relations

$$\lambda(f,g) = (Af,g) = (f,Ag) = \overline{\mu}(f,g)$$

it follows that $(f,g) = 0$. In particular, if $\lambda \neq \overline{\lambda}$ and $Af = \lambda f$, then $(f,f) = 0$.

From this it is clear that in each of the open upper and lower half planes there are not more than $\varkappa$ eigenvalues of A, as in the contrary case the corresponding eigenvectors would form a neutral subspace of dimension greater than $\varkappa$, a contradiction to Lemma 1.2.

Theorem 6.1. Let A be a closed Hermitian operator in $\Pi_\varkappa$ ($\overline{D(A)} = \Pi_\varkappa$). Then the points ζ of the open upper half plane are points of regular type of A, and with the possible exception of at most $\varkappa$ points the defect number $\dim(R(A - \zeta I))^\perp$ is constant there (= n_+). The exceptional points ζ coincide with the eigenvalues of A in the open upper half plane, and for these we have

$$\dim(R(A - \zeta I))^\perp = n_+ + r_\zeta, \tag{6.5}$$

where r_ζ is the algebraic multiplicity[1]) of the eigenvalue ζ.

Proof. The representation (6.4) of A and Theorem 9.4 of [17a] imply that the points ζ, $\mathrm{Im}\,\zeta > 0$, with the possible exception of an at most countable number, are of regular type, $\dim(R(A - \zeta I))^\perp$ is constant and equals $\dim(R(A_0 - \zeta I))^\perp$, and the last statement of the theorem holds. On the other hand, it was shown before the theorem that A has at most $\varkappa$ eigenvalues in the open upper half plane.

Note 1. A corresponding statement holds for the lower half plane $\mathrm{Im}\,\zeta < 0$ with a defect number n_- which can be different from n_+.

Note 2. If A is selfadjoint, then $n_+ = n_- = 0$.

Indeed, suppose e.g. $n_+ > 0$. Then for infinitely many points ζ, $\mathrm{Im}\,\zeta > 0$, there exist vectors $h_\zeta \neq 0$ such that $(Af - \zeta f, h_\zeta) = 0$ for all $f \in D(A)$. This implies $h_\zeta \in D(A^*) = D(A)$ and $Ah_\zeta = A^*h_\zeta = \bar{\zeta}h$, a contradiction.

6.3. Definition 6.3. A linear operator V, which maps its domain D(V) onto R(V), is called[2]) isometric if

$$(Vx, Vy) = (x, y) \quad (x, y \in D(V)) . \tag{6.6}$$

Similarly as in case of Hermitian operators one can prove that the (formally weaker) requirement $(Vx, Vx) = (x, x)$ $(x \in D(V))$ is equivalent to (6.6).

---

[1])This is the dimension of the root manifold S_ζ.

[2])If we require in addition that the scalar product does not degenerate on D(V), then from the linearity of V and (6.6) it follows automatically that the mapping of D(V) onto R(V) is one-to-one. In this case the scalar product does not degenerate on R(V) as well.

Conversely, the one-to-oneness of the mapping $VD(V) = R(V)$ together with (6.6) implies that V is linear if only the scalar product on D(V) does not degenerate.

The operator V in Definition 6.3 is not necessarily invertible. Consider e.g. the linear operator V defined on the linear span D(V) of some neutral vector $x \neq \theta$ by $Vx = \theta$. However, if V is invertible, then evidently also V^{-1} is an isometric operator.

In a Hilbert space an isometric operator is bounded. The following examples show that this is not true even in Π_1.

Example 6.1. Let the space Π_1 be infinite-dimensional. Choose an arbitrary neutral vector $(\theta \neq)$ $e \in \Pi_1$ and let $M = E^{\perp}$ be the orthogonal complement of the linear span E of e. Then M is an infinite-dimensional nonpositive subspace which contains the one-dimensional isotropic subspace E, and we have (see Theorem 3.3) $M = M_1 \oplus E$, where M_1 is a negative subspace.

We choose a discontinuous linear functional F on M with the property $F(e) = 1$ and define on M the linear operator V:

$$Vx = F(x)e - x \quad (x \in M) .$$

Then $Ve = 0$ and $(Vx,Vy) = (x,y)$. But as F is not continuous, there exists a sequence (x_n), $x_n \in M$, $\|x_n\| = 1$, $n = 1,2,\dots$, such that $\lim_{n\to\infty}|F(x_n)| = \infty$. It follows that

$$\lim_{n\to\infty}\|Vx_n\| = \lim_{n\to\infty}\|F(x_n)e - x_n\| \geq \lim_{n\to\infty}(\|F(x_n)e\| - \|x_n\|) =$$

$$= \lim_{n\to\infty}|F(x_n)|\,\|e\| - 1 = \infty .$$

Note. The set of zeros of the discontinuous functional F is a linear manifold L ($\subset M$) which is dense in M ($\bar{L} = M$). As $e \in L$, the scalar product does not degenerate on L, but it degenerates on $M = \bar{L}$ (cf. 3.4).

The isometric operator V in Example 6.1 is not invertible ($Ve = \theta$). But it needs only a slight modification to show that even an invertible isometric operator may be discontinuous.

Example 6.2. Let us restrict the operator V from Example 6.1 to the subspace M_1:

$$V_o = V|M_1, \quad V_o x = F(x)e - x \quad (x \in M_1) .$$

The functional F is evidently discontinuous also on M_1 ($\dim M/M_1 = 1$), and hence the same is true for V_o. Now we extend V_o again on the whole set M, defining the linear operator $\widetilde{V}$ as follows:

$$\widetilde{V}x = \begin{cases} V_o x, & x \in M_1 \\ \lambda e, & x = \lambda e \end{cases} .$$

It is easily seen that the discontinuous operator V maps M one-to-one and isometrically onto itself.

A continuous isometric operator in Π_κ evidently has a continuous closure which is again an isometric operator. The next theorem shows that the converse statement is also true.

Theorem 6.2. A closed (or closable) isometric operator V in Π_κ is continuous.

Proof. Assume that the closed isometric operator V is discontinuous. Then there exists a sequence (y_n), $y_n \in D(V)$, $\|Vy_n\| = 1$, $n = 1,2,\ldots$, and $\|y_n\| \to 0$ $(n \to \infty)$, and we have

$$(Vy_n, Vy_n) = (y_n, y_n) \to 0 \quad (n \to \infty) . \tag{6.7}$$

But V is also closed as an operator from D(V) (lying in the Hilbert space $\overline{D(V)}$) into the Hilbert space $\overline{R(V)}$ (see e.g. [20], Chapter III, § 5); that is (ibid. Theorem 5.29), there exists an adjoint operator V' (with respect to the H-scalar product in $\overline{D(V)}$ and $\overline{R(V)}$), densely defined in $\overline{R(V)}$ and with values in $\overline{D(V)}$. Therefore

$$[Vy_n, z] = [y_n, V'z] \to 0 \quad (n \to \infty)$$

for all $z \in D(V')$, which implies $[Vy_n, z] \to 0$ $(n \to \infty)$ for all $z \in \overline{R(V)}$ because $\overline{R(V)} = \overline{D(V')}$ and $\|Vy_n\| = 1$. On the other hand, obviously $[Vy_n, z] = 0$, $n = 1,2,\ldots$, for all $z \in R(V)^{[\perp]}$, and we get $(Vy_n, z) \to 0$ $(n \to \infty)$ for all $z \in \Pi_\kappa$. Together with (6.7), by Theorem 2.4, (i), this implies $Vy_n \to \theta$ $(n \to \infty)$, a contradiction.

Note. Theorem 6.2 is valid (see [18f]) for a larger class of operators in Π_κ.

Theorem 6.3. If V is an isometric operator in Π_κ and the scalar product does not degenerate on $\overline{R(V)}$, then V is continuous.

Proof. According to Theorem 6.2 it is sufficient to prove that V is closable, i.e., the closure of its graph is again a graph. Assuming the contrary, there exists a sequence $x_n \in D(V)$ $(n = 1,2,\ldots)$ such that

$$x_n \to \theta, \quad Vx_n \to y_o \neq \theta \quad (n \to \infty) .$$

But $y_0 \in \overline{R(V)}$ and for all $x \in D(V)$

$$(Vx,y_0) = \lim_{n\to\infty}(Vx,Vx_n) = \lim_{n\to\infty}(x,x_n) = (x,\theta) = 0 ,$$

i.e., y_0 is an isotropic vector of $\overline{R(V)}$, a contradiction.

Corollary 1. If V is an isometric operator in $\Pi_\varkappa$ and the scalar product does not degenerate on $\overline{D(V)}$ and $\overline{R(V)}$, then V and V^{-1} are continuous.

Corollary 2. If V is an isometric operator in $\Pi_\varkappa$ and D(V) contains a $\varkappa$-dimensional positive subspace, then V and V^{-1} are continuous.

Indeed, in this case also R(V) contains a $\varkappa$-dimensional positive subspace; hence on $\overline{D(V)}$ and $\overline{R(V)}$ the scalar product does not degenerate (see the note after Definition 3.1).

Definition 6.4. The isometric operator U is called unitary if $D(U) = R(U) = \Pi_\varkappa$.

Thus a unitary operator maps the whole of $\Pi_\varkappa$ one-to-one onto itself, preserving the scalar product of any pair of vectors.

If $D(U) = \Pi_\varkappa$, but $R(U) \neq \Pi_\varkappa$, then the isometric operator U is called semi-unitary.

If U is a unitary (semi-unitary) operator in $\Pi_\varkappa$, then, clearly, U^{-1} is a unitary (isometric) operator, and from Corollary 2 of Theorem 6.3 it follows that they are both continuous[1]. In addition for unitary U the relation $(Ux,y) = (x,U^{-1}y)$, which is valid for any $x,y \in \Pi_\varkappa$, shows that

$$U^* = U^{-1}, \quad (U^{-1})^* = (U^*)^{-1} = U . \tag{6.8}$$

Hence a linear operator U in $\Pi_\varkappa$ with $D(U) = \Pi_\varkappa$ is unitary if and only if

$$U^*U = UU^* = I ,$$

and semi-unitary if and only if

$$U^*U = I, \quad UU^* \neq I .$$

An example of a unitary operator in $\Pi_\varkappa$ is the operator J (see 3.1), since from properties 1° and 3° of this operator it follows that $(Jx,Jy) = (x,y)$.

[1]) Pontrjagin [35] called a unitary operator an automorphism, and made the initial requirement that it should be linear and continuous. But the linearity is a consequence of the assumption that the mapping U from $\Pi_\varkappa$ onto $\Pi_\varkappa$ is one-to-one and preserves the scalar product, and the continuity follows as above.

6.4. Obviously, in general the classes of selfadjoint and unitary operators are widely different from one another. It is enough to mention that selfadjoint operators are not always bounded. Nevertheless there are analogies between the properties of selfadjoint (Hermitian) and unitary (isometric, respectively) operators in $\Pi_\varkappa$, the origin of which is explained in § 8. As one example of this analogy we give two theorems on the spectra of unitary and selfadjoint operators.

Theorem 6.4. The spectrum of a selfadjoint operator A in $\Pi_\varkappa$ is located symmetrically with respect to the real axis. Every nonreal number is either an eigenvalue or a regular point of A, and there are at most $\varkappa$ eigenvalues of A in the open upper half plane.

Proof. If ζ, $\operatorname{Im}\zeta \neq 0$, is a regular point of A, then $A - \zeta I$ maps D(A) one-to-one onto $\Pi_\varkappa$, and $(A - \zeta I)^{-1}$ is continuous. From (5.3), (5.1') and Definition 6.2 we get

$$((A - \zeta I)^{-1})^* = ((A - \zeta I)^*)^{-1} = (A - \bar{\zeta} I)^{-1} ,$$

therefore $\bar{\zeta}$ is a regular point of A. As the spectrum is the complement of the set of regular points, the first statement follows. As for the second statement see Note 2 after Theorem 6.1.

Theorem 6.5. The spectrum of a unitary operator U in $\Pi_\varkappa$ is located symmetrically with respect to the unit circle. Every point λ, $|\lambda| \neq 1$, is either an eigenvalue or a regular point of U, and there are at most $\varkappa$ eigenvalues in the open unit disc.

Proof. Clearly, $\lambda = 0$ is a regular point of U. Let λ ($\lambda \neq 0$) be a regular point of U. From (5.3), (5.1') and (6.8) we get that

$$((U - \lambda I)^{-1})^* = ((U - \lambda I)^*)^{-1} = (U^{-1} - \bar{\lambda} I)^{-1} ;$$

that is, $\bar{\lambda}$ is a regular point of U^{-1}. On the other hand,

$$U^{-1} - \bar{\lambda} I = - \bar{\lambda} U^{-1} (U - \bar{\lambda}^{-1} I) ,$$

from which it is evident that $\bar{\lambda}^{-1}$ is a regular point of U, and the first statement follows.

We shall not prove the second statement here. Later (§ 8) it will become clear that it is an immediate consequence of Theorem 6.4 and the Cayley transformation (as is also the first statement of Theorem 6.5).

6.5. In the preceding sections unitary and semi-unitary operators in Π_κ have been considered as particular cases of isometric operators. They are also particular cases of other classes of operators introduced in this section. Here and in the rest of § 6 we always suppose $\dim \Pi_\kappa > \kappa$.

Definition 6.5. A continuous linear operator T in Π_κ with $D(T) = \Pi_\kappa$ is called a plus-operator if

$$(x,x) \geqq 0 \quad \text{implies} \quad (Tx,Tx) \geqq 0 \quad (x \in \Pi_\kappa) .$$

A simple lemma is important in this context. We quote it from [26] (see also [9]; in similar but less complete form it was proved independently by several authors ([12], [14], [27])).

Lemma 6.1. Suppose on a linear space E two bilinear metrics[1]) (x,y) and $(x,y)_1$ are given and (x,y), $x,y \in E$, is indefinite; that is, there are vectors x_+, $x_- \in E$ such that $(x_+,x_+) > 0$ and $(x_-,x_-) < 0$. If $(x,x) = 0$ implies $(x,x)_1 \geqq 0$, then

a) $\mu \equiv \inf\{(x,x)_1 \mid (x,x) = 1\} > -\infty$,

b) $(x,x)_1 \geqq \mu(x,x)$ for all $x \in E$.

Proof. First we shall show that for every pair $y,z \in E$ with $(y,y) < 0$ and $(z,z) > 0$ we have

$$(y,y)_1 \big/ (y,y) \leqq (z,z)_1 \big/ (z,z) \ . \tag{6.9}$$

Indeed, otherwise there would be elements $y,z \in E$ such that $(y,y) = -1$, $(z,z) = 1$ and $-(y,y)_1 > (z,z)_1$. Setting $x = \varepsilon y + z$, where $|\varepsilon| = 1$, we obtain

$$(x,x) = 2\,\mathrm{Re}\{\varepsilon(y,z)\}, \qquad (x,x) < 2\,\mathrm{Re}\{\varepsilon(y,z)_1\} \ .$$

Thus for suitable ε we have $(x,x) = 0$ and $(x,x)_1 < 0$, in contradiction with the hypothesis.

Now (6.9) immediately implies a). The inequality in b) follows from the definition of μ if $(x,x) > 0$, and from (6.9) if $(x,x) < 0$. Finally, if $(x,x) = 0$, it holds by the condition of the lemma.

Lemma 6.1 has an immediate consequence.

---

[1]) See axiom II in 1.1.

Theorem 6.6. Let T be a plus-operator in Π_κ. Then

$$(Tx,Tx) \geqq \mu(T)\,(x,x) \quad \text{for all } x \in \Pi_\kappa ,$$

where

$$\mu(T) = \inf_{(x,x)=1} (Tx,Tx) \quad (\geqq 0) . \tag{6.10}$$

To prove this statement it is sufficient to apply Lemma 6.1 to the bilinear metrics (x,y) and $(x,y)_1 = (Tx,Ty)$.

If $\mu(T) = 0$, the range of T is nonnegative. Hence, by Lemma 1.2 it is an at most κ-dimensional subspace. Therefore we shall concentrate our attention on the case $\mu(T) > 0$.

The plus-operator T is said to be <u>strict</u> if $\mu(T) > 0$.

A similar class of operators is introduced by the following definition.

Definition 6.6. A continuous linear operator T in Π_κ with $D(T) = \Pi_\kappa$ is called <u>nondecreasing</u> if

$$(x,x) \geqq 0 \text{ implies } (Tx,Tx) \geqq (x,x) \quad (x \in \Pi_\kappa) . \tag{6.11}$$

Evidently, unitary and semi-unitary operators in Π_κ are nondecreasing. Moreover, a nondecreasing operator T is a strict plus-operator (with $\mu(T) = 1$). On the other hand, if T is a strict plus-operator then $\mu(T)^{-1/2}T$ is nondecreasing and, moreover, expansive (see the following Definition 6.7).

Definition 6.7. A continuous linear operator T in Π_κ with $D(T) = \Pi_\kappa$ is called <u>expansive</u> if

$$(Tx,Tx) \geqq (x,x) \quad \text{for all } x \in \Pi_\kappa .$$

A more special but nevertheless important class of operators is introduced by the following definition.

Definition 6.8. A continuous linear operator T in Π_κ with $D(T) = \Pi_\kappa$ is called <u>uniformly expansive</u> if

$$(Tx,Tx) \geqq (x,x) + \delta^2\|x\|^2 \quad \text{for some } \delta > 0 \text{ and all } x \in \Pi_\kappa .$$

It is easy to see that the set of uniformly expansive operators is open in the Banach algebra $\mathfrak{B}(\Pi_\kappa)$ of all linear continuous operators in Π_κ.

It follows from the Definitions 6.7 and 6.8 that the set $\mathfrak{E}$ ($\mathfrak{E}^0$) of all expansive (uniformly expansive, respectively) operators forms a semigroup with respect to the ordinary multiplication of operators. Moreover, the identity

$$(UVx,UVx) - (x,x) = (UVx,UVx) - (Vx,Vx) + (Vx,Vx) - (x,x) \qquad (x \in \Pi_\varkappa)$$

shows that if $U \in \mathfrak{E}$ and $V \in \mathfrak{E}^0$, then $UV \in \mathfrak{E}^0$.

It turns out that $\mathfrak{E}$ is the closure of $\mathfrak{E}^0$ $(\subset \mathfrak{E})$ in the topology of $\mathfrak{B}(\Pi_\varkappa)$, i.e., given $U \in \mathfrak{E}$ and $\varepsilon > 0$ there exists a $V \in \mathfrak{E}^0$ such that $\|U - V\| < \varepsilon$.

To prove this, note that for every δ $(0 < \delta < 1)$ the operator S_δ:

$$S_\delta x = \sqrt{1 + \delta^2}x_+ + \sqrt{1 - \delta^2}x_- \quad (x \in \Pi_\varkappa = \Pi_+ + \Pi_-, x = x_+ - x_-, x_\pm \in \Pi_\pm) \tag{6.12}$$

is uniformly expansive:

$$(S_\delta x, S_\delta x) = (1 + \delta^2)\|x_+\|^2 - (1 - \delta^2)\|x_-\|^2 = (x,x) + \delta^2\|x\|^2 .$$

Then $V = US_\delta \in \mathfrak{E}^0$, and for an arbitrary $\varepsilon > 0$ we have

$$\|U - V\| = \|U - US_\delta\| \leqq \|U\|\|I - S_\delta\| < \varepsilon$$

if only δ is sufficiently small.

A series of spectral properties of operators $U \in \mathfrak{E}$ or $U \in \mathfrak{E}^0$ will be established in Chapter III.

In connection with these classes of operators the following remarks are of some interest. We suppose $\dim \Pi_\varkappa > \varkappa$. Let T be a continuous linear operator in $\Pi_\varkappa$, $D(T) = \Pi_\varkappa$, for which there exists a real number μ such that

$$(Tx,Tx) \geqq \mu(x,x) \quad \text{for all} \quad x \in \Pi_\varkappa . \tag{6.13}$$

Then two real numbers $\mu_+(T)$ and $\mu_-(T)$ can be defined by

$$\mu_+(T) = \inf_{(x,x)=1} (Tx,Tx) , \quad \mu_-(T) = \sup_{(x,x)=-1} (-(Tx,Tx)) .$$

Evidently, $\mu_-(T) \leqq \mu \leqq \mu_+(T)$, and, conversely, for each number μ with this property relation (6.13) holds. If T is a plus-operator, the number $\mu_+(T)$ is the same as $\mu(T)$ in (6.10), and hence $\mu_+(T) \geqq 0$. If T is expansive, it follows that $\mu_-(T) \leqq 1 \leqq \mu_+(T)$.

If $\dim \Pi_\varkappa > 2\varkappa$, the number μ in (6.13) is nonnegative. [1])Indeed,

---

[1]) The operator $T = \begin{bmatrix} 0 & 1 \\ 0 & 0 \end{bmatrix}$ in the two-dimensional space Π_1 of all vectors $x = \begin{bmatrix} \xi_1 \\ \xi_2 \end{bmatrix}$ with scalar product $(x,x) = |\xi_1|^2 - |\xi_2|^2$ shows that the condition about the dimension is essential.

otherwise we would have $(Tx,Tx) \geqq |\mu|\cdot|(x,x)| > 0$ for $(x,x) < 0$, which would imply that T maps a $(\varkappa+1)$-dimensional negative subspace one-to-one onto an at most $\varkappa$-dimensional positive subspace, a contradiction.

The following fact is evident:

1°. If T satisfies (6.13) for two different values μ, say μ and μ', $\mu < \mu'$, then $\mu_-(T) \leqq \mu < \mu' \leqq \mu_+(T)$.

2°. If T satisfies (6.13) and $\mu_-(T) < \mu_+(T)$, then for each μ, $\mu_-(T) < \mu < \mu_+(T)$ there exists a $\delta(\mu) > 0$ such that

$$(Tx,Tx) \geqq \mu(x,x) + \delta(\mu)\|x\|^2 \quad \text{for all } x \in \Pi_\varkappa \, . \tag{6.14}$$

Conversely, if (6.14) holds with some $\delta(\mu) > 0$, then $\mu_-(T) \leqq \mu - \delta(\mu) < \mu \leqq \mu_+(T)$.

Indeed, the selfadjoint operator $A = T^*T$ has the property

$$((A - \mu I)x,x) \geqq 0 \quad \text{for all } x \in \Pi_\varkappa,\ \mu \in [\mu_-,\mu_+], \tag{6.15}$$

where $\mu_\pm := \mu_\pm(T)$. It follows that

$$((A-\mu_+I)x,x) \geqq ((A-\mu I)x,x) \geqq ((A-\mu_-I)x,x) \geqq 0 \quad \text{if } (x,x) \leqq 0 \, ,$$

$$((A-\mu_-I)x,x) \geqq ((A-\mu I)x,x) \geqq ((A-\mu_+I)x,x) \geqq 0 \quad \text{if } (x,x) \geqq 0 \, .$$

If (6.14) were false, we could find a sequence $(x_n) \in \Pi_\varkappa$ with $\|x_n\| = 1$ and $((A - \mu I)x_n,x_n) \to 0$ $(n \to \infty)$. This implies, by Cauchy's inequality, that $(A - \mu I)x_n \to 0$ $(n \to \infty)$. On the other hand, consider the subsequences $(x_n^\pm)$ of (x_n) such that $(x_n^\pm,x_n^\pm) \gtreqless 0$. If, e.g., (x_n^+) has an infinite number of elements, it follows that $(A - \mu_+I)x_n^+ \to \theta$ $(n \to \infty)$; hence $(\mu - \mu_+)x_n^+ \to \theta$, a contradiction. Therefore we have for $\mu \in (\mu_-,\mu_+)$

$$((A - \mu I)x,x) \geqq \delta(\mu)\|x\|^2 \quad (x \in \Pi_\varkappa) \tag{6.16}$$

with some $\delta(\mu) > 0$. Conversely, (6.16) implies (6.15) because of the inequality $\|x\|^2 \geqq -(x,x)$.

As immediate consequences of (6.14) we have

3°. If T satisfies (6.13) and $\mu_-(T) < \mu_+(T)$, then $0 \notin (\mu_-(T),\mu_+(T))$.

Otherwise we would have $(Tx,Tx) \geqq \delta(0)\|x\|^2$ for all $x \in \Pi_\varkappa$ and some $\delta(0) > 0$, which is impossible since $\dim \Pi_\varkappa > \varkappa$.

4°. If T satisfies (6.13) and $\mu_-(T) < \mu_+(T)$, then the interval $(\mu_-(T),\mu_+(T))$ belongs to $\rho(T^*T)$.

Indeed, assume $\mu \in (\mu_-(T),\mu_+(T)) \cap \sigma(A)$, $A = T^*T$. Then there exists a sequence (x_n), $\|x_n\| = 1$, such that $(A - \mu I)x_n \to \theta$ $(n \to \infty)$, which is impossible because of (6.14) or (6.16).

These statements $1^0 - 4^0$ can be applied to a plus-operator T. For example, we find from 3^0 that for a strict plus-operator T we have not only $\mu_+(T) > 0$, but also $\mu_-(T) \geqq 0$. Further properties of plus-operators will be established in 11.1 - 11.4 and 12.6.

Definition 6.9. A continuous linear operator T in $\Pi_\varkappa$ with $D(T) = \Pi_\varkappa$ is called a B-plus-operator[1]) if

$$(x,x) \geqq 0, \; x \neq 0, \; \text{implies } (Tx,Tx) > 0 \; . \tag{6.17}$$

Evidently, every uniformly expansive operator is a B-plus-operator. We leave it to the reader to construct an example of a B-plus-operator which is not uniformly expansive.

Lemma 6.2. Every B-plus-operator is a strict plus-operator.

Proof. Evidently, T is a plus-operator. Assume $\mu(T) = 0$. Then the range of T is a $\varkappa$-dimensional positive subspace. If we choose this subspace as Π_+ in a decomposition (1.1) then in the matrix representation $T = (T_{ij})$, $i,j = 1,2$, we have $T_{21} = T_{22} = 0$. The assumption $\mu(T) = 0$ implies that there exist sequences $(x_n) \subset \Pi_+$, $(y_n) \subset \Pi_-$ such that $\|x_n\|^2 = \|y_n\|^2 + 1$ and $\|T_{11}x_n + T_{12}y_n\|^2 \to 0$ $(n \to \infty)$. As $\|x_n\| \geqq 1$, the sequence $(x_n/\|x_n\|)$ converges to some $x_0 \neq \theta$ and $(y_n/\|x_n\|)$ converges weakly to some y_0. Then $T_{11}x_0 + T_{12}y_0 = \lim_{n \to \infty} \|x_n\|^{-1}(T_{11}x_n + T_{12}y_n) = \theta$. On the other hand we have $\|x_0\| = 1$, $\|y_0\| \leqq \varliminf_{n \to \infty} \|x_n\|^{-1}\|y_n\| \leqq 1$, hence $\|x_0\|^2 - \|y_0\|^2 \geqq 0$. This is impossible by (6.17). The lemma is proved.

Every unitary operator U in $\Pi_\varkappa$ is a strict plus-operator but not a B-plus-operator.

Lemma 6.3. If T is a B-plus-operator in $\Pi_\varkappa$, then T^* is also a B-plus-operator.[2])

Proof. Let L be an arbitrary $\varkappa$-dimensional nonnegative subspace. Consider all $x \in \Pi_\varkappa$, $x \neq 0$, such that $(T^*L,x) = \{0\}$, i.e., $(L,Tx) = \{0\}$.

[1])This class of operators was introduced by M.L. Brodskiĭ in [10].

[2])Similar affirmations about expansive, uniformly expansive and (strict) plus-operators are valid (see Lemma 11.3 below).

According to Lemma 1.2, $(Tx,Tx) \leq 0$; therefore $(x,x) < 0$ (see (6.17)). Hence T^*L is a positive $\varkappa$-dimensional subspace, and so $(T^*x,T^*x) > 0$ for $x \in L$, $x \neq \theta$. Elementary arguments show that each x with $(x,x) \geq 0$ may be imbedded in a $\varkappa$-dimensional nonnegative subspace L (cf. a more general fact established below in Lemma 10.2). Thus for such x ($\neq \theta$) we have $(T^*x,T^*x) > 0$ and T^* is a B-plus-operator.

In concluding this section it is worth observing that if in Definitions 6.6 - 6.9 we replace the requirements for the operator T to be continuous and $D(T) = \Pi_\varkappa$ by the following: "T is closed and densely defined", we do not obtain larger classes of operators. This follows from [10].

§ 7. Root manifolds and elementary divisors of isometric operators in $\Pi_\varkappa$

7.1. A main difference between isometric (including unitary) and Hermitian (including selfadjoint) operators in $\Pi_\varkappa$ and those of the same names in Hilbert space is that in $\Pi_\varkappa$ they can have eigenvalues which, for the isometric operators, need not have modulus one and, for the Hermitian operators, need not be real. Both sorts of operators can have nonsimple elementary divisors.

As L.S. Pontrjagin showed [35], the number of nonreal eigenvalues for Hermitian operators in $\Pi_\varkappa$ and the number and orders of the elementary divisors are connected in a definite way with the number $\varkappa$.

Analogous theorems hold for isometric operators in $\Pi_\varkappa$, as is shown below by the same methods as were used in [35].

7.2. Theorem 7.1. If λ and μ are eigenvalues of the isometric operator V, and $\lambda \neq \bar{\mu}^{-1}$, then the corresponding root manifolds S_λ and S_μ of the operator V are orthogonal to one another. In particular, if $\lambda = \mu$ and $|\lambda| \neq 1$, then S_λ is a neutral subspace.

Proof. We must show that if $\lambda \neq \bar{\mu}^{-1}$ then from

$$(V - \lambda I)^r x = \theta \ , \quad (V - \mu I)^s y = \theta \tag{7.1}$$

it follows that $(x,y) = 0$ (r,s natural numbers). For $r = s = 0$ (i.e., $r + s = 0$) this assertion is trivial. Now let us suppose that it has been proved for any two indices with sum less than $r + s$. Write

$$(V - \lambda I)x = x', \quad (V - \mu I)y = y' \ . \tag{7.2}$$

Then from (7.1) and (7.2) we get that

$$(V - \lambda I)^{r-1}x' = \theta \,, \quad (V - \mu I)^{s}y = \theta \,,$$
$$(V - \lambda I)^{r}x = \theta \,, \quad (V - \mu I)^{s-1}y' = \theta \tag{7.3}$$

(we suppose that $r \geqq 1$ and $s \geqq 1$, since otherwise either $x = \theta$ or $y = \theta$).

By the induction assumption we get from (7.3) that $(x',y) = (x,y') = 0$, whence, using (7.2), we obtain

$$(x,y) = (Vx,Vy) = (\lambda x + x', \mu y + y') = \lambda\bar{\mu}(x,y) \,.$$

The last equation, because of $\lambda\bar{\mu} \neq 1$, gives $(x,y) = 0$.

Corollary 1. A root manifold S_λ of an isometric operator corresponding to a number λ with $|\lambda| \neq 1$ has a dimension not greater than $\varkappa$.

Corollary 2. An isometric operator cannot have more than $\varkappa$ different eigenvalues whose moduli exceed one. The same is true of the eigenvalues whose moduli are less than one. Thus, the total number of eigenvalues which are not equal to one in modulus cannot be greater than $2\varkappa$.

A further refinement of this statement will be formulated as a corollary to Theorem 7.3.

7.3. Now we consider root manifolds of an isometric operator V corresponding to eigenvalues of modulus one.

Definition 7.1. We say that an eigenvalue λ of a linear operator T corresponds to a nonsimple elementary divisor if there are a vector x and a natural number $r > 1$ such that

$$(T - \lambda I)x \neq \theta \,, \quad (T - \lambda I)^{r}x = \theta \,.$$

Theorem 7.2. If V is a closed isometric operator in $\Pi_\varkappa$ and ε $(|\varepsilon| = 1)$ is an eigenvalue of V, then the root manifold S_ε can be put in the form of an orthogonal direct sum of subspaces invariant for V: $S_\varepsilon = S \oplus S'$, where S' consists only of eigenvectors of V and S is a finite-dimensional subspace. Thus V can have nonsimple elementary divisors only in the finite-dimensional subspace S.

Proof. If S_ε is finite-dimensional, it is sufficient to put $S = S_\varepsilon$. Suppose that S_ε is infinite-dimensional.

a) Without loss of generality we can suppose that $\varepsilon = 1$, which corresponds to changing from V to the operator $\varepsilon^{-1}V$. We write accordingly $S_\varepsilon = S_1$.

If p is an arbitrary natural number, then by S_1^p we mean the set of all vectors $x \in S_1$ for which $(V - I)^p x = \theta$:

$$S_1^p = \{x \mid (V - I)^p x = \theta\} .$$

It is clear that

$$S_1^p \subset S_1^{p+1} \quad (p = 1,2,\ldots) .$$

In particular, S_1^1 is a linear manifold consisting of all fixed points of V: $S_1^1 = \{x \mid Vx = x\}$. Thus, as V is closed, S_1^1 is a closed linear manifold, i.e. a subspace.

In order to show that S_1 is also a subspace, it is enough to show that $\dim S_1/S_1^1 < \infty$. This is shown below (see section d) of this proof).

b) We show first that $\dim S_1^{p+1}/S_1^p < \varkappa \quad (p = 1,2,\ldots)$.

In fact, let R be a linear manifold in S_1^{p+1} which is such that the intersection of it with S_1^p contains only the vector zero:

$$R \subset S_1^{p+1} , \quad S_1^p \cap R = \{\theta\} . \tag{7.4}$$

Let $g \in R$ and $h \in S_1^1$; then

$$(h,(V - I)^p g) = (Vh,(V - I)(V - I)^{p-1}g) = (h - Vh,(V - I)^{p-1}g)$$
$$= (\theta,(V - I)^{p-1}g) = 0 ,$$

i.e., $S_1^1 \perp (V - I)^p R$. But, since $R \subset S_1^{p+1}$, we have $(V - I)^p R \subset S_1^1$, i.e., $(V - I)^p R$ is orthogonal to itself, and so its dimension cannot exceed $\varkappa$ (see Lemma 1.2).

We note now that the linear transformation $(V - I)^p$ of the linear manifold R into S_1^1 is one-to-one, since from $g_1, g_2 \in R$ and $(V - I)^p g_1 = (V - I)^p g_2$ it follows that $(V - I)^p (g_1 - g_2) = \theta$, $g_1 - g_2 \in S_1^p$, $g_1 - g_2 \in R$, and so from (7.4) $g_1 - g_2 = \theta$, i.e. $g_1 = g_2$. Consequently, the dimension of R is equal to the dimension of $(V - I)^p R$ and does not exceed $\varkappa$.

c) Let q be a natural number such that $S_1^{q-1} \neq S_1^q$; then, as was shown above,

$$S_1^q = S_1^{q-1} \dotplus R^q , \tag{7.5}$$

where the dimension of R^q is not greater than $\varkappa$. Obviously, R^q consists of the vector θ and those vectors $x \neq \theta$ for which

$$(V - I)^q x = \theta , \quad (V - I)^{q-1} x \neq \theta .$$

Each vector $g = (V - I)x$ $(x \in R^q)$ has the properties $(V - I)^{q-1}g = \theta$ and $(V - I)^{q-2}g \neq \theta$, i.e., $g \in S_1^{q-1}$, but $g \notin S_1^{q-2}$.

In this way,

$$(V - I)R^q \cap S_1^{q-2} = \{\theta\}, \qquad (V - I)R^q \subset S_1^{q-1} . \tag{7.6}$$

Since $S_1^{q-2} \subset S_1^{q-1}$, (7.6) shows that S_1^{q-2} is a proper part of S_1^{q-1}, i.e., once again

$$S_1^{q-1} = S_1^{q-2} \dot{+} R^{q-1} ,$$

where R^{q-1} is a subspace of positive dimension not exceeding $\varkappa$. Besides this, the first of the relations (7.6) gives

$$(V - I)R^q \subset R^{q-1} .$$

Extending these arguments further, we obtain a chain of formulae

$$S_1^p = S_1^{p-1} \dot{+} R^p, \qquad (V - I)R^p \subset R^{p-1} \qquad (p = 2,3,\ldots,q) , \tag{7.7}$$

where we put $R^1 = S_1^1$. Adding the equations (7.7) for p ranging from 2 to q, we get

$$S_1^q = R^1 \dot{+} \ldots \dot{+} R^q .$$

d) We introduce the notation[1])

$$(V - I)^p R^{2p} = Q \qquad (p = 1,2,\ldots,[q/2]). \tag{7.8}$$

From (7.8) and (7.7) we have $Q^p \subset R^p$. We write out the direct sum

$$P^q = Q^1 \dot{+} Q^2 \dot{+} \ldots \dot{+} Q^{[q/2]} \tag{7.9}$$

and show that $Q^j \perp Q^k$ for $j,k \leq [q/2]$.

In fact, if $j \leq k \leq [q/2]$, then for any $f \in Q^j$ and $g \in Q^k$ we have from (7.8) and (7.5) that

$$f = (V - I)^j x, \quad g = (V - I)^k y \quad (x \in R^{2j},\ y \in R^{2k}) ,$$
$$(f,g) = ((V - I)^j x,\ (V - I)^k y) = ((V^{-1} - I)^k\ (V - I)^j x,y) =$$
$$= ((-1)^k V^{-k}(V - I)^{j+k}x,y) = 0 ,$$

since from $j \leq k$ it follows that $j + k \geq 2j$, whence $(V - I)^{j+k}x = \theta$ $(x \in R^{2j} \subset S_1^{2j}$.

It follows from the proof that the scalar product is identically zero

1)The symbol [a] denotes the largest integer less than or equal to a.

on P^q, and this implies that the dimension of P^q is not greater than $\varkappa$, and, in particular, that $[q/2] \leqq \varkappa$. The last fact follows from the observation, that no subspace Q^p $(p = 1,2,\dots,[q/2])$ reduces to zero, since the mapping (7.8) is one-to-one (see section b) of the proof) and $R^1,\dots,R^q$ do not reduce to zero.

e) Thus, $q \leqq 2\varkappa + 1$, i.e., the sequence $S_1^1, S_1^2, \dots$ after some number becomes stable (consists of sets which coincide). If q is this number $(q \leqq 2\varkappa + 1)$, then we have (see c))

$$S_1 = S_1^q = R^1 \dotplus R^2 \dotplus \dots \dotplus R^q = S_1^1 \dotplus R^2 \dotplus R^3 \dotplus \dots \dotplus R^q , \tag{7.10}$$

and since the dimensions of the R^p $(p = 2,\dots,q)$ are finite, (see section b)), the dimension $\dim S_1/S_1^1$ is finite (it does not exceed $(q - 1)\varkappa \leqq 2\varkappa^2$). As S_1^1 is closed (see section a)), S_1 is closed, i.e., is a subspace of $\Pi_\varkappa$.

f) Now we consider the decomposition

$$S_1^1 = S_o \oplus M \tag{7.11}$$

of the subspace S_1^1 according to Theorem 3.3. Here S_o is the isotropic subspace of S_1^1, and the scalar product does not degenerate on the subspace M. Therefore M may be decomposed as the orthogonal sum of a negative subspace S' and a positive subspace Q: $M = S' \oplus Q$ (note that S' as well as S_1 and S_1^1 are infinite-dimensional). Consequently (see Theorem 3.2)

$$\Pi_\varkappa = S' \oplus \Pi'_\varkappa , \tag{7.12}$$

where $\Pi'_\varkappa$ is again a space of type $\Pi_\varkappa$.

Since $S' \subset S_1^1$, S' consists only of fixed vectors for the operator V; in particular $VS' = S'$, whence it follows that $\Pi'_\varkappa$ is also invariant under V, i. e., $x \in \Pi'_\varkappa \cap D(V)$ implies $Vx \in \Pi'_\varkappa$.

We write

$$S = S_1 \cap \Pi'_\varkappa . \tag{7.13}$$

Clearly, it follows from $S' \subset S_1^1 \subset S_1$, (7.12) and (7.13) that

$$S_1 = S \oplus S' . \tag{7.14}$$

From the relations $VS_1 \subset S_1$ and $VS' = S'$ the invariance of the subspace S follows: $VS \subset S$.

Since, in the negative subspace S', V is simply the identical operator $(V = I)$, the operator V can have nonsimple elementary divisors, corresponding to $\varepsilon = 1$, only in S.

g) Introducing the notation

$$S^p = \{x \in S \mid (V - I)^p x = \theta\} \quad (p = 1,2,\dots) ,$$

we again find that dim S^{p+1}/S^p is finite and does not exceed $\varkappa$ (it coincides with dim S_1^{p+1}/S_1^p since $S_1^p = S^p \oplus S'$).

Since all the arguments of sections a) - e) can be repeated for the chain $S^1 \subset S^2 \subset \dots$, we again find that dim $S/S^1 < \infty$; but from (7.11) we have

$$S^1 \subset S_1^1 = S' \oplus Q , \tag{7.15}$$

where Q is finite-dimensional. On the other hand, since $S^1 \subset S$ and $S \cap S' = \{\theta\}$ (see (7.14)), it follows from (7.15) and the relation $S^1 \cap S' = \{\theta\}$, that S^1 is finite dimensional. But dim $S/S^1 < \infty$, so that S is also finite-dimensional, and this completes the proof of the theorem.

7.4. From Theorem 7.2 it follows that on the finite-dimensional subspace S the isometric operator V can be expressed as an ordinary matrix, and the concept of elementary divisors gets its usual meaning. In particular, we can speak of the order of the elementary divisors: non-simple elementary divisors are divisors of order higher than one.

The following theorem shows that the number and the orders of the non-simple elementary divisors of V are connected with the number $\varkappa$.

Theorem 7.3. Let S_ε ($|\varepsilon| = 1$) be a root manifold of the closed isometric operator V in $\Pi_\varkappa$ and $S_\varepsilon = S \oplus S'$ its decomposition according to Theorem 7.2. If V has elementary divisors in S with orders $d_1,\dots,d_r$, then

$$\rho(\varepsilon) = \sum_{j=1}^{r} \left[{}^{d_j}/2\right] \leq \varkappa . \tag{7.16}$$

Proof. Again we suppose that $\varepsilon = 1$ and $S_\varepsilon = S_1$.

For the proof we use the notation and arguments which were used in the proof of the preceding theorem (sections a) - e)), relating them not to S_1 but to the finite-dimensional space S.

We denote the dimension of R^p by r^p:

$$\dim R^p = \dim S^p/S^{p-1} = r^p \quad (p = 1,\dots,q) .$$

From (7.8) we conclude that the dimension of Q^p is equal to r^{2p} ($p = 1,\dots,[{}^q/2]$). The dimension ρ of the subspace $P^q \subset S$ (see (7.9)) is equal to

$$\rho = r^2 + r^4 + \ldots + r^{2[q/2]} .$$

Consider a vector x^p such that $x^p \in R^p$ but $x^p \notin (V - I)R^{p+1}$ (if such a vector exists). It is easy to verify that for any $p = 2, \ldots, q$ the vectors

$$x^p, (V - I)x^p, \ldots, (V - I)^{p-1}x^p \tag{7.17}$$

are linearly independent.

In fact, if

$$a_0x^p + a_1(V - I)x^p + \ldots + a_{p-1}(V - I)^{p-1}x^p = \theta , \tag{7.18}$$

then, applying the operator $(V - I)^{p-1}$, we get (since $x^p \in R^p \subset S^p$) that

$$a_0(V - I)^{p-1}x^p = \theta .$$

If $a_0 \neq 0$, then $x^p \in S^{p-1}$. But $S^p = S^{p-1} \dotplus R^p$, i.e., $x^p = \theta$, which was assumed not to be the case. Thus $a_0 = 0$. Applying the operators $(V - I)^{p-2}, (V - I)^{p-3}, \ldots, V - I, I$ to both sides of (7.18), we show in an analogous way that

$$a_0 = a_1 = \ldots = a_{p-1} = 0 .$$

Now consider[1])

$$N_{x^p} = \text{l. s. } \{x^p, (V - I)x^p, \ldots, (V - I)^{p-1}x^p\} . \tag{7.19}$$

N_{x^p} is invariant with respect to V and is a cyclic subspace corresponding to an elementary divisor of order p.

Let $x^p, y^p, \ldots, z^p$ be a maximal system of vectors in R^p linearly independent with respect to $(V - I)R^{p+1}$.[2]) On each of these we can construct a cyclic subspace of the form (7.19), and each of these subspaces $N_{x^p}, N_{y^p}, \ldots, N_{z^p}$ corresponds to an elementary divisor of order p.

We shall now show that the subspace S decomposes into a direct sum of cyclic subspaces $N_{x^p}, N_{y^p}, \ldots, N_{z^p}$ $(p = 1, \ldots, q)$.

---

[1])The letters l.s. here and in future denote <u>linear span</u>, c.l.s. denote <u>closed linear span</u>.

[2])That is to say, they are such that no linear combination of these vectors with nonzero coefficients is in $(V - I)R^{p+1}$.

In fact, $S = S^q$ (see (7.10)) or $S = R^1 \dotplus \ldots \dotplus R^q$, where $(V - I)R^{p+1} \subset R^p$ $(p = 1,\ldots,q-1)$. Commencing with R^q, let us consider its basis $x^q, y^q, \ldots, z^q$ consisting of $\sigma^q = r^q$ vectors. Construct $N_{x^q}, N_{y^q}, \ldots, N_{z^q}$ and note that the linear independence of the systems of vectors $x^q, (V - I)x^q, \ldots, (V - I)^{q-1}x^q$; $y^q, (V - I)y^q, \ldots, (V - I)^{q-1}y^q$; $z^q, (V - I)z^q, \ldots, (V - I)^{q-1}z^q$ can be proved in exactly the same way as for the system (7.17).

Now from R^{q-1} we select all the vectors which are linearly independent with respect to $(V - I)R^q$ (the vectors from $(V - I)R^q$ are already in $N_{x^q}, N_{y^q}, \ldots, N_{z^q}$). Let these vectors be

$$x^{q-1},\ y^{q-1},\ \ldots,\ z^{q-1}\ . \tag{7.20}$$

Clearly $R^{q-1} = \text{l.s.}\left\{x^{q-1}, y^{q-1}, \ldots, z^{q-1}\right\} \dotplus (V - I)R^q$. The number σ^{q-1} of vectors (7.20) is equal to $r^{q-1} - r^q$. We construct the cyclic subspaces $N_{x^{q-1}}, N_{y^{q-1}}, \ldots, N_{z^{q-1}}$, etc. Proceeding in this way up to R^1, we exhaust all of S.

We see further that the number σ^p of elementary divisors of the operator V of order p is equal to the number of cyclic subspaces $N_{x^p}, N_{y^p}, \ldots, N_{z^p}$, i.e.,[1])

$$\sigma^p = r^p - r^{p+1} \quad (p = 2,3,\ldots,q;\ r^{q+1} = 0)\ . \tag{7.21}$$

From this we get

$$r^p = \sigma^p + r^{p+1} = \sigma^p + \sigma^{p+1} + r^{p+2} = \ldots = \sigma^p + \sigma^{p+1} + \ldots + \sigma^q \quad (p = 2,3,\ldots,q).$$

$$\begin{aligned}
r^2 &= \sigma^2 + \sigma^3 + \sigma^4 + \sigma^5 + \sigma^6 + \sigma^7 + \ldots + \sigma^q\ ,\\
r^4 &= \sigma^4 + \sigma^5 + \sigma^6 + \sigma^7 + \ldots + \sigma^q\ ,\\
r^6 &= \sigma^6 + \sigma^7 + \ldots + \sigma^q\ ,\\
&\ldots\ldots\ldots\ldots\ldots\ldots\ldots\ldots\ldots\ldots\\
r^{2[q/2]} &= (\sigma^{2[q/2]}) + \sigma^q
\end{aligned}$$

$$\rho = \sum_{p=1}^{[q/2]} r^{2p} = \sigma^2 + \sigma^3 + 2\sigma^4 + 2\sigma^5 + 3\sigma^6 + 3\sigma^7 + \ldots + [q/2]\sigma^q = \sum_{p=2}^{q} [p/2]\sigma^p\ .$$

1) The number $\sigma^1 = r^1 - r^2$ does not give the number of all the elementary divisors of order one, but only the number of them in S.

The letter ρ here denotes the dimension of P^q (see (7.9)), which, as we have shown, does not exceed $\varkappa$. On the other hand, as σ^p is the number of elementary divisors of order p we see that (see the formulation of the theorem)

$$\varkappa \geqq \rho = \sum_{p=2}^{q} [p/2]\sigma^p = \sum_{j=1}^{r} [^{d_j}/2] ,$$

which completes the proof.

Note. (7.21) shows that the number σ^p of elementary divisors of order p ($p = 2,\ldots,q$) is completely determined by the indices $r^2,\ldots,r^q$ which do not depend on the particular construction of the finite-dimensional space S, as carried out in section f) of the proof of Theorem 7.2. These indices, as we have seen, are completely determined by the linear manifold S_ε itself, and so by V and the number ε.

In this way, the number of nonsimple elementary divisors of a given order corresponding to an eigenvalue ε is an invariant of the operator V and ε, and this justifies the notation $\rho(\varepsilon)$ for the sum (7.16).

Corollary. Let $\lambda_1,\ldots,\lambda_n$ be a set of eigenvalues of the isometric operator V in $\Pi_\varkappa$, and let this set not contain two numbers symmetric with respect to $|z| = 1$. For $|\lambda| \neq 1$ let $\rho(\lambda)$ denote the dimension of the root manifold S_λ, and for $|\lambda| = 1$ the sum (7.16).

Then $\rho(\lambda_1) + \ldots + \rho(\lambda_n) \leqq \varkappa$.

For the proof it is enough to consider the subspaces P_λ which coincide with S_λ for $|\lambda| \neq 1$ and with P^q (see (7.9)) for $|\lambda| = 1$

In both cases $P_{\lambda_j} \perp P_{\lambda_k}$ ($j \gtrless k$), as follows from Theorem 7.1 and section d) of the proof of Theorem 7.2. Hence, in the orthogonal direct sum

$$P = P_{\lambda_1} \oplus \ldots \oplus P_{\lambda_n}$$

the scalar product is identically zero, consequently the dimension of P does not exceed $\varkappa$:

$$\rho(\lambda_1) + \ldots + \rho(\lambda_n) \leqq \varkappa .$$

§ 8. The connection between Hermitian and isometric operators (Cayley transformation)

8.1. As in Hilbert space, the Hermitian and isometric operators in $\Pi_\varkappa$ are closely connected. This connection can be realized by the so-called

Cayley transformation.

Let A be an Hermitian operator in Π_κ. We choose a nonreal number ζ which is not an eigenvalue of A (see the remarks before Theorem 6.1) and define an operator U by the formulae

$$g = Af - \zeta f, \; Ug = Af - \bar{\zeta} f \quad (f \in D(A)) \tag{8.1}$$

or, symbolically,

$$U = (A - \bar{\zeta}I)(A - \zeta I)^{-1} = I + (\zeta - \bar{\zeta})(A - \zeta I)^{-1} . \tag{8.2}$$

The operator U is called the <u>Cayley transform</u> of the operator A.

The formulae (8.1) define a vector f uniquely for each element g from the domain $D(U) = \{Af - \zeta f \mid f \in D(A)\}$ and for this f they then define a vector Ug. It is easily seen that the operator U is linear. In addition it preserves the scalar product: In fact,

$$(Ug,Ug) = (Af - \bar{\zeta}f, Af - \bar{\zeta}f) = (Af,Af) - 2\mathrm{Re}\{\bar{\zeta}(f,Af)\} + |\zeta|^2(f,f) ,$$
$$(g,g) = (Af - \zeta f, Af - \zeta f) = (Af,Af) - 2\mathrm{Re}\{\bar{\zeta}(Af,f)\} + |\zeta|^2(f,f) ,$$

and consequently $(Ug,Ug) = (g,g)$.

Thus we have proved the first part of the following proposition:

1°. The operator U defined by (8.1) is isometric, and $1 \notin \sigma_p(U)$. Moreover, $\overline{(U - I)D(U)} = \Pi_\kappa$.

The last assertion follows immediately from (8.1), since $(U - I)D(U) = D(A)$.

2°. The operator U defined by (8.1) is invertible if and only if $\bar{\zeta} \notin \sigma_p(A)$.

This follows directly from the first equality in (8.2).

3°. Let the Hermitian operator A in Π_κ be closed. Then its Cayley transform U is also closed.

Indeed, as A and $(A - \zeta I)^{-1}$ are both closed, the assertion follows from (8.2) (see the last term there).

The next proposition is a simple consequence of 2°, 3° and Theorem 6.2.

4°. If A is a closed Hermitian operator in Π_κ, both the Cayley transform U from (8.2) and its inverse U^{-1} are continuous; D(U) and R(U) are closed.

5°. If A is a selfadjoint operator in Π_κ ($A = A^*$), then its Cayley transform U is a unitary operator.

Indeed, in this case $D(U) = \{Af - \zeta f \mid f \in D(A)\} = \Pi_\varkappa$ and $R(U) = \{Af - \bar{\zeta} f \mid f \in D(A)\} = \Pi_\varkappa$ (see Note 2 to Theorem 6.1).

In concluding this section we note the following: If for a given operator A we wish to construct an operator U such that U is isometric and $\{Ug - \varepsilon g \mid g \in D(U)\}$ is dense in $\Pi_\varkappa$, where ε ($|\varepsilon| = 1$) is a number given in advance, then we should take the Cayley transform in the form

$$g = \varepsilon^{-1}(Af - \zeta f), \; Ug = Af - \bar{\zeta} f \tag{8.3}$$

or

$$U = \varepsilon(A - \bar{\zeta} I)(A - \zeta I)^{-1} .$$

<u>8.2.</u> Now let us invert the transformation (8.1) - (8.2).

Elementary calculations give

$$f = \frac{1}{\zeta - \bar{\zeta}}(Ug - g), \; Af = \frac{1}{\zeta - \bar{\zeta}}(\zeta Ug - \bar{\zeta} g) \qquad (g \in D(U)) . \tag{8.4}$$

If we consider ζ in (8.4) as an arbitrary nonreal number and U as a preasssigned isometric operator with $(D(A) =)\ (U - I)D(U)$ satisfying the condition $\overline{D(A)} = \Pi_\varkappa$, then the operator A given by (8.4) or

$$A = (\zeta U - \bar{\zeta} I)(U - I)^{-1} = \zeta I + (\zeta - \bar{\zeta})(U - I)^{-1} \tag{8.5}$$

is called the <u>Cayley transform</u> of the operator U.

6°. The Cayley transform A of the isometric operator U is a Hermitian operator. If U is closed, then A is also closed.

Indeed, $\overline{D(A)} = \Pi_\varkappa$, and for all $f \in D(A)$ according to (8.4) the number

$$(Af,f) = \frac{1}{|\zeta - \bar{\zeta}|^2}(\zeta Ug - \bar{\zeta} g, Ug - g) =$$

$$= \frac{1}{|\zeta - \bar{\zeta}|^2}[\zeta(g,g) - \zeta(Ug,g) - \bar{\zeta}(g,Ug) + \bar{\zeta}(g,g)]$$

is real. The second statement of 6° follows from (8.5).

7°. Let U be a unitary operator in $\Pi_\varkappa$, and $1 \notin \sigma_p(U)$. Then the linear manifold $(U - I)D(U)$ $(= D(A))$ is dense in $\Pi_\varkappa$, the Cayley transform A of U is a selfadjoint operator, and both ζ and $\bar{\zeta}$ are regular points of A.

Indeed, if for all $g \in \Pi_\varkappa$ and some $f_0 \in \Pi_\varkappa$ we have

$$0 = (Ug - g, f_0) = (g, U^{-1}f_0 - f_0) ,$$

then $U^{-1}f_0 = f_0$, i.e., $Uf_0 = f_0$ or $f_0 = \theta$. Thus $\overline{D(A)} = \Pi_\varkappa$. Now (8.5) with regard of (5.1'), (5.3) and (6.8) gives

$$A^* = \bar{\zeta}I - (\bar{\zeta} - \zeta)(U^{-1} - I)^{-1} = \bar{\zeta}I + (\bar{\zeta} - \zeta)U(I - U)^{-1} =$$

$$= [\bar{\zeta}(I - U) + (\bar{\zeta} - \zeta)U](I - U)^{-1} = (\bar{\zeta}I - \zeta U)(I - U)^{-1} =$$

$$= [\zeta(I - U) - (\zeta - \bar{\zeta})I](I - U)^{-1} = \zeta I + (\zeta - \bar{\zeta})(U - I)^{-1} = A .$$

The last statement (about ζ and $\bar{\zeta}$) follows immediately from (8.5) and Theorem 6.4.

Similarly to 8.1, the Cayley transform of an isometric operator U may be constructed starting not necessarily from $\varepsilon = 1$ but from any point ε, $|\varepsilon| = 1$, with the property $\overline{(U - \varepsilon I)D(U)} = \Pi_\varkappa$. In this more general case the Cayley transform A has a form

$$A = (\zeta U - \bar{\zeta}\varepsilon I)(U - \varepsilon I)^{-1}$$

or, more explicitly,

$$f = \frac{1}{\zeta - \bar{\zeta}} (Ug - \varepsilon g), \quad Af = \frac{1}{\zeta - \bar{\zeta}} (\zeta Ug - \varepsilon\bar{\zeta}g) \quad (g \in D(U)) . \tag{8.6}$$

The results of the sections 8.1 - 2 may be summarized (omitting some details) as follows:

Theorem 8.1. Let A be a Hermitian operator in $\Pi_\varkappa$, $\zeta \neq \bar{\zeta}$ and $|\varepsilon| = 1$. If $\zeta \notin \sigma_p(A)$, then the Cayley transform of A, i.e.,

$$U = \varepsilon(A - \bar{\zeta}I)(A - \zeta I)^{-1} , \tag{8.3'}$$

is an isometric operator, and $\overline{(U - \varepsilon I)D(U)} = \Pi_\varkappa$. In addition, the inverse transformation has the form

$$A = (\zeta U - \bar{\zeta}\varepsilon I)(U - \varepsilon I)^{-1} . \tag{8.6'}$$

The operator U is invertible if and only if $\bar{\zeta} \notin \sigma_p(A)$, it is closed if and only if A is closed, and it is a unitary operator if and only if A is selfadjoint.

Conversely, if U is isometric and $\overline{(U - \varepsilon I)D(U)} = \Pi_\varkappa$, then its Cayley transform (8.6') is a Hermitian operator in $\Pi_\varkappa$, with $\zeta \notin \sigma_p(A)$, and (8.3') holds.

In case of a separable space $\Pi_\varkappa$ and a unitary[1]) operator U there is at least one ε, $|\varepsilon| = 1$, such that the condition $\overline{(U - \varepsilon I)D(U)} = \Pi_\varkappa$ is

[1])This condition can be weakened. For example, it is sufficient that U is isometric and at least one of the sets D(U) or R(U) coincides with $\Pi_\varkappa$.

satisfied. This can be shown using the facts that two eigenvectors of U belonging to different eigenvalues on the unit circle are orthogonal (Theorem 7.1) and, on the other hand, in a separable space $\Pi_\varkappa$ an arbitrary orthonormal system is at most countable (see, e.g., [9]).
We mention, in this connection, that for a given continuous linear operator T in a nonseparable space $\Pi_\varkappa$ there exists a separable space $\Pi'_\varkappa$ (of the same type), such that

$$\Pi_\varkappa = \Pi'_\varkappa \oplus \Pi'_0$$

and the corresponding matrix representation of T is

$$T = \begin{bmatrix} T' & 0 \\ 0 & T'_0 \end{bmatrix} .$$

Evidently, Π'_0 is a Hilbert space with respect to the scalar product $-(x,y)$ $(x,y \in \Pi'_0)$. Therefore the investigation of T in the nonseparable space $\Pi_\varkappa$ can often be reduced to that of T' in the separable space $\Pi'_\varkappa$ (and of T'_0 in the Hilbert space Π'_0).
To give one choice of $\Pi'_\varkappa$ we fix some basis $e_1,\dots,e_\varkappa$ in Π_+ and consider the elements

$$T^{n_1}(T^*)^{n_2}\,T^{n_3}\,\dots\,T^{n_{2p-1}}(T^*)^{n_{2p}}\,e_k, \qquad k = 1,\dots,\varkappa, \quad 0 \leqq n_p < \infty, \quad p = 1,2,\dots .$$

Their closed linear span is a separable subspace of $\Pi_\varkappa$ containing Π_+, which is invariant under T and T*. Hence it has all the properties of the space $\Pi'_\varkappa$ above.

8.3. The mutually inverse Cayley transformations

$$A = (\zeta U - \bar{\zeta}\varepsilon I)(U - \varepsilon I)^{-1}$$

$$U = \varepsilon(A - \bar{\zeta}I)(A - \zeta I)^{-1} \qquad (\zeta \neq \bar{\zeta};\ |\varepsilon| = 1)$$

correspond to the fractional linear transformations

$$\lambda = \frac{\zeta\upsilon - \bar{\zeta}\varepsilon}{\upsilon - \varepsilon} \tag{8.7}$$

and

$$\upsilon = \varepsilon\,\frac{\lambda - \bar{\zeta}}{\lambda - \zeta}, \tag{8.8}$$

respectively.
The following proposition is then immediately evident from (8.3) and (8.6').

Theorem 8.2. If υ ($\upsilon \neq \varepsilon$) is an eigenvalue of U, and $Ug = \upsilon g$ ($g \neq \theta$), then g is an eigenvector of the operator A, corresponding to the eigenvalue λ (see (8.7)).

Conversely, if $Af = \lambda f$ ($\lambda \neq \zeta$, $f \neq \theta$), then f is an eigenvector of U, corresponding to the eigenvalue υ (see (8.8)).

This proposition has the following generalization.

Theorem 8.2'. Let the operators A and U be connected by the mutually inverse Cayley transformations

$$g = \varepsilon^{-1}(Af - \zeta f), \quad Ug = Af - \bar{\zeta} f , \tag{8.3}$$

$$f = \frac{1}{\zeta - \bar{\zeta}} (Ug - \varepsilon g), \quad Af = \frac{1}{\zeta - \bar{\zeta}} (\zeta Ug - \bar{\zeta}\varepsilon g) . \tag{8.6}$$

Then any finite-dimensional invariant subspace M of the operator U is invariant for A, and conversely.

Proof. If $g \in M$, then $Ug \in M$. The formulae (8.6) then give

$$f = \frac{1}{\zeta - \bar{\zeta}} (Ug - \varepsilon g) \in M, \quad Af = \frac{1}{\zeta - \bar{\zeta}} (\zeta Ug - \varepsilon\bar{\zeta} g) \in M . \tag{8.9}$$

If $g_1, \ldots, g_p$ ($p < \infty$) is a basis for the subspace M, then the vectors $f_1, \ldots, f_p$ defined by

$$g_j = \varepsilon^{-1}(Af_j - \zeta f_j), \quad (j = 1,2,\ldots,p) ,$$

also form a basis for M, as is easily seen, and the relations (8.9) applied to the vectors $f = f_j$ ($j = 1,\ldots,p$) show that $Af_j \in M$ ($j = 1,\ldots,p$) i.e., $AM \subset M$.

The second part of the theorem has a completely analogous proof.

8.4. Now we study the behavior of the root manifolds of operators under the Cayley transformation.

Let A and U be connected by (8.3) and (8.6). Then the following two propositions hold:

a) If for some $x \in D(A)$ the vector $(A - \lambda I)^r x$ exists, where r is a natural number, then

$$(A - \lambda I)^r x = (U - \upsilon I)^r y ,$$

where

$$y = \alpha^r (A - \zeta I)^r x \in D(U); \quad \alpha = \frac{1}{\varepsilon - \upsilon}, \quad \upsilon = \varepsilon \frac{\lambda - \bar{\zeta}}{\lambda - \zeta} .$$

In fact, for $r = 1$, since $x \in D(A)$, we can put $y_1 = \varepsilon^{-1}(A - \zeta I)x$. Then on the basis of (8.3) we deduce that

$$(U - \upsilon I)y_1 = Ax - \zeta x - \upsilon\varepsilon^{-1} Ax + \upsilon\varepsilon^{-1} \zeta x =$$

$$= (1 - \upsilon\varepsilon^{-1})Ax - (\bar{\zeta} - \upsilon\varepsilon^{-1}\zeta)x = (1 - \upsilon\varepsilon^{-1})(A - \lambda I)x = \varepsilon^{-1}(\varepsilon - \upsilon)(A - \lambda I)x .$$

If we now put $y = \varepsilon(\varepsilon - \upsilon)^{-1}y_1 = \alpha(A - \zeta I)x$, then

$$(U - \upsilon I)y = (A - \lambda I)x .$$

This implies that the proposition holds for $r = 1$.

Let the proposition now be established for all exponents which do not exceed $r - 1$. Then

$$(A - \lambda I)^r x = (A - \lambda I)^{r-1}(A - \lambda I)x$$
$$= \alpha^{r-1}(U - \upsilon I)^{r-1}(A - \zeta I)^{r-1}(A - \lambda I)x$$
$$= \alpha^{r-1}(U - \upsilon I)^{r-1}(A - \lambda I)(A - \zeta I)^{r-1}x$$
$$= \alpha^r(U - \upsilon I)^{r-1}(U - \upsilon I)(A - \zeta I)^r x = (U - \upsilon I)^r y .$$

b) If $y \in D(U)$ and $(U - \upsilon I)^r y$ exists, then with

$$x = \beta^r(U - \varepsilon I)^r y \in D(A), \quad \beta = \frac{1}{\zeta - \lambda}, \quad \lambda = \frac{\zeta\upsilon - \bar{\zeta}\varepsilon}{\upsilon - \varepsilon}$$

we have

$$(U - \upsilon I)^r y = (A - \lambda I)^r x .$$

The proof is analogous to the preceding one.

It is easy to verify that the formulae

$$x = \beta^r(U - \varepsilon I)^r y, \quad y = \alpha^r(A - \zeta I)^r x \tag{8.10}$$

in propositions a) and b) are mutually inverse.

From a) and b) it follows that the relations

$$(A - \lambda I)^{r-1}x \neq \theta, \quad (A - \lambda I)^r x = \theta \tag{8.11}$$

imply

$$(U - \upsilon I)^{r-1}y \neq \theta, \quad (U - \upsilon I)^r y = \theta$$

and conversely; here λ and υ are related by (8.7) and (8.8), and x and y by (8.10).

It is evident from (8.11) that $M = \text{l.s.}\ \{x, Ax, \ldots, A^r x\}$ is a finite-dimensional invariant (cyclic) subspace for A, and so (by Theorem 8.2') for U. On the other hand, the second of the formulae (8.10) shows that $y \in M$. Hence

$$N = \text{l.s.}\ \{y, Uy, \ldots, U^r y\} \subset M \ .$$

In exactly the same way we can show that $M \subset N$; hence $M = N$. This proves that the cyclic subspaces and the corresponding root manifolds of the operators A and U are identical.[1])

8.5. Using the results obtained, we can carry over (with consequential changes) all the theorems and corollaries of § 7 to closed Hermitian operators in $\Pi_\varkappa$. The reader can find without trouble the modified forms of Theorems 7.1 - 3 and their corollaries, in which the role of the unit circle is taken over by the real axis.[2])

In particular, the estimate (7.16) remains valid. In [35] L.S. Pontrjagin did not say if this estimate is exact. The following example, suggested by an analogous construction due to A.I. Mal'cev ([31], p. 388) shows that it is so.

We shall construct a Hermitian operator A in an n-dimensional ($n = 2\varkappa + 1$) space $\Pi_\varkappa$, which has, for example, the elementary divisor $(\lambda - a)^{2\varkappa+1}$ (a real).

To build the space $\Pi_\varkappa$ we define in the n-dimensional coordinate space a bilinear (indefinite) metric by means of a nondegenerate Hermitian form (see 4.4)

$$(x,x) = -\xi_1\overline{\xi}_{2\varkappa+1} - \xi_2\overline{\xi}_{2\varkappa} - \ldots - \xi_{\varkappa+1}\overline{\xi}_{\varkappa+1} - \ldots - \xi_{2\varkappa}\overline{\xi}_2 - \xi_{2\varkappa+1}\overline{\xi}_1,$$

which has, as can be easily verified, $\varkappa$ positive squares.

The scalar product of two vectors in $\Pi_\varkappa$:

---

[1])For unitary and Hermitian operators in finite-dimensional spaces $\Pi_\varkappa$ this fact is well-known (see, e.g., A.I. Mal'cev [31], p. 416).

[2])Conversely, if the analogues of Theorems 7.1 - 3, which were established by Pontrjagin [35], are used as a starting point, then the theorems themselves (for isometric operators) can be obtained by means of the Cayley transformation. We mention that Pontrjagin [35] proved the analogues of Theorems 7.2 and 7.3 for a closed Hermitian operator. In [4a] it was shown that they also hold for a nonclosed Hermitian operator and even in slightly more general spaces than $\Pi_\varkappa$. In the same way, also in Theorems 7.2 and 7.3 it would be enough to suppose that the isometric operator V has a closure.

$$x = (\xi_1, \xi_2, \dots, \xi_{2\varkappa+1}), \qquad y = (\eta_1, \eta_2, \dots, \eta_{2\varkappa+1})$$

is given by

$$(x,y) = -\xi_1\bar{\eta}_{2\varkappa+1} - \xi_2\bar{\eta}_{2\varkappa} - \dots - \xi_{2\varkappa+1}\bar{\eta}_1 .$$

For the basis vectors

$$e_1 = (1,0,0,\dots,0),\ e_2 = (0,1,0,\dots,0),\dots,\ e_{2\varkappa+1} = (0,0,0,\dots,1)$$

the Gram matrix $G = \|(e_j,e_k)\|_{j,k=1}^{2\varkappa+1}$ has the form

$$G = \begin{bmatrix} 0 & 0 & \dots & 0 & 0 & -1 \\ 0 & 0 & \dots & 0 & -1 & 0 \\ 0 & 0 & \dots & -1 & 0 & 0 \\ \dots & \dots & \dots & \dots & \dots & \dots \\ 0 & -1 & \dots & 0 & 0 & 0 \\ -1 & 0 & \dots & 0 & 0 & 0 \end{bmatrix} .$$

We now define the operator A in $\Pi_\varkappa$ by the formulae

$$Ae_1 = ae_1,\ Ae_2 = e_1 + ae_2,\ Ae_3 = e_2 + ae_3,\dots,\ Ae_{2\varkappa+1} = e_{2\varkappa} + ae_{2\varkappa+1} .$$

The structure of the matrix G ensures that A is Hermitian. On the other hand, by its construction A has one elementary divisor $(\lambda - a)^{2\varkappa+1}$ in $\Pi_\varkappa$.

8.6. Finally we shall establish the connection between the spectra of the operators A and U.

Theorem 8.3. Let the closed Hermitian operator A and the isometric operator U in $\Pi_\varkappa$ be connected by the relations (8.3) and (8.6), i.e., by the Cayley transformations

$$A = (\zeta U - \bar{\zeta}\varepsilon I)(U - \varepsilon I)^{-1}, \quad U = \varepsilon(A - \bar{\zeta}I)(A - \zeta I)^{-1} \quad (\zeta \neq \bar{\zeta},\ |\varepsilon| = 1) .$$

Then any regular point υ ($\upsilon \neq \varepsilon$) of the operator U corresponds to a regular point

$$\lambda = \frac{\zeta\upsilon - \bar{\zeta}\varepsilon}{\upsilon - \varepsilon} \tag{8.7}$$

of the operator A, and conversely any regular point λ ($\lambda \neq \zeta$) of A corresponds to a regular point

$$\upsilon = \varepsilon\,\frac{\lambda - \bar{\zeta}}{\lambda - \zeta} \tag{8.8}$$

of U.

Proof. a) Let υ ($\upsilon \neq \varepsilon$) be a regular point of U. This implies that for any $h \in \Pi_\varkappa$ there is a unique vector g in D(U) such that

$$\upsilon g - Ug = h , \tag{8.12}$$

and there is a constant $C > 0$ satisfying the condition $\|g\| \leqq C\|h\|$ for all h and g connected by (8.12).

If now we take a vector f given by the first of the formulae (8.6), then g and Ug are expressed by f and Af by means of (8.3).

Consequently, (8.12) can be rewritten in the form

$$\upsilon\varepsilon^{-1}(Af - \zeta f) - (Af - \bar{\zeta} f) = h$$

or

$$Af_0 - \frac{\zeta\upsilon - \bar{\zeta}\varepsilon}{\upsilon - \varepsilon} f_0 = h, \quad Af_0 - \lambda f_0 = h , \tag{8.13}$$

where

$$f_0 = \frac{\upsilon - \varepsilon}{\varepsilon} f, \quad \lambda = \frac{\zeta\upsilon - \bar{\zeta}\varepsilon}{\upsilon - \varepsilon} .$$

Thus for any $h \in \Pi_\varkappa$ there is a vector $f_0 \in D(A)$ which is such that (8.13) holds. This vector f_0 is uniquely defined, since otherwise from (8.13) we would get that λ is an eigenvalue of A, i.e., υ is an eigenvalue of U (Theorem 8.2).

Besides this, we have

$$\|f_0\| = \left|\frac{\upsilon - \varepsilon}{\varepsilon}\right| \cdot \|f\| = C_1 \frac{1}{|\zeta - \bar{\zeta}|}\|Ug - \varepsilon g\| \leqq \frac{C_1}{|\zeta - \bar{\zeta}|}(\|Ug\| + \|g\|) \leqq$$

$$\leqq \frac{C_1}{|\zeta - \bar{\zeta}|} (\|U\| + 1)\|g\| \leqq \frac{C_1}{|\zeta - \bar{\zeta}|} (\|U\| + 1)C\|h\| \leqq C_2\|h\| ,$$

where

$$C_1 = \left|\frac{\upsilon - \varepsilon}{\varepsilon}\right|, \quad C_2 = \frac{C_1}{|\zeta - \bar{\zeta}|} (\|U\| + 1)C .$$

Since the constant C_2 does not depend on h, this shows that λ is a regular point of A.

b) Conversely, let λ ($\lambda \neq \zeta$) be a regular point of A. This implies that for any $h \in \Pi_\varkappa$ there is a unique vector $f \in D(A)$ which is such that

$$Af - \lambda f = h, \quad \|f\| \leqq C\|h\| , \tag{8.14}$$

where the constant C does not depend on h.

In particular, there is a vector $f_0 \in D(A)$ such that

$$Af_0 - \lambda f_0 = -\frac{\lambda - \zeta}{\zeta - \bar{\zeta}}\, h \, . \tag{8.15}$$

Now write $g_0 = \varepsilon^{-1}(Af_0 - \zeta f_0) \in D(U)$; then we get from (8.15), (8.6) and (8.8) that

$$(\zeta - \bar{\zeta})(\lambda f_0 - Af_0) = (\lambda - \zeta)h, \quad \lambda(Ug_0 - \varepsilon g_0) - (\zeta Ug_0 - \bar{\zeta}\varepsilon g_0) = (\lambda - \zeta)h \, ,$$

$$(\lambda - \zeta)Ug_0 - (\lambda - \bar{\zeta})\varepsilon g_0 = (\lambda - \zeta)h, \quad Ug_0 - \varepsilon\,\frac{\lambda - \bar{\zeta}}{\lambda - \zeta}\, g_0 = h, \quad Ug_0 - \upsilon g_0 = h \, .$$

In this way we can find for an arbitrary vector $h \in \Pi_\varkappa$ a vector $g_0 \in D(U)$ such that

$$Ug_0 - \upsilon g_0 = h \, . \tag{8.16}$$

The relation (8.16) defines g_0 uniquely, since in the opposite case υ would be an eigenvalue of U, i.e., λ would be an eigenvalue of A.

Besides this, from (8.15) and (8.14) we have

$$\|g_0\| = \|Af_0 - \zeta f_0\| = \|Af_0 - \lambda f_0 + (\lambda - \zeta)f_0\| \leqq$$

$$\leqq \|Af_0 - \lambda f_0\| + |\lambda - \zeta|\,\|f_0\| \leqq \left|\frac{\lambda - \zeta}{\zeta - \bar{\zeta}}\right| \|h\| + |\lambda - \zeta|\, C\|h\| = C_1\|h\| \, ,$$

where $C_1 = |\lambda - \zeta|\left(\frac{1}{|\zeta - \bar{\zeta}|} + C\right)$ does not depend on h.

The theorem is completely proved.

<u>Corollary.</u> If υ ($\upsilon \neq \varepsilon$) is a point of the spectrum of U, then λ (see (8.7)) is a point of the spectrum of A. Conversely, if λ ($\lambda \neq \zeta$) is a point of the spectrum of A, then υ (see (8.8)) is a point of the spectrum of U.

The proof follows at once from Theorem 8.3 if we note that from the relation $\zeta \neq \bar{\zeta}$ it follows that

$$\lambda = \frac{\zeta\upsilon - \bar{\zeta}\varepsilon}{\upsilon - \varepsilon} \neq \zeta \quad \text{(for all } \upsilon) \quad \text{and} \quad \upsilon = \varepsilon\,\frac{\lambda - \bar{\zeta}}{\lambda - \zeta} \neq \varepsilon \quad \text{(for all } \lambda) \, .$$

Theorem 8.3 and its Corollary may be sharpened as follows:

<u>Theorem 8.3'.</u> If under the conditions of Theorem 8.3 υ ($\upsilon \neq \varepsilon$) is a point of the residual spectrum of U: $\upsilon \in \sigma_r(U)$, then

$$\lambda = \frac{\zeta\upsilon - \bar{\zeta}\varepsilon}{\upsilon - \varepsilon} \in \sigma_r(A) .$$

Conversely, if (ζ $\neq$) $\lambda \in \sigma_r(A)$, then

$$\upsilon = \varepsilon \frac{\lambda - \bar{\zeta}}{\lambda - \zeta} \in \sigma_r(U) .$$

This follows from Theorem 8.2 and the formula $(A - \lambda I)D(A) = (U - \upsilon I)D(U)$, where λ and υ are connected by (8.7) and (8.8).

Corollary. If υ ($\upsilon \neq \varepsilon$) is a point of the continuous spectrum of U: $\upsilon \in \sigma_c(U)$, then

$$\lambda = \frac{\zeta\upsilon - \bar{\zeta}\varepsilon}{\upsilon - \varepsilon} \in \sigma_c(A) .$$

Conversely, if (ζ $\neq$) $\lambda \in \sigma_c(A)$, then

$$\upsilon = \varepsilon \frac{\lambda - \bar{\zeta}}{\lambda - \zeta} \in \sigma_c(U) .$$

This follows from a comparison of Theorems 8.3, 8.2 and 8.3'.

§ 9. Extensions of isometric and Hermitian operators

9.1. Let V be a continuous and invertible isometric operator in $\Pi_\varkappa$ with closed domain D(V). Then V^{-1} is a closed isometric operator in $\Pi_\varkappa$, which is also continuous by Theorem 6.2, therefore $D(V^{-1}) = R(V)$ is also closed. The defect index of V is by definition the pair (n_0, n_1), where n_0 and n_1 are the so-called defect numbers, i.e.

$$n_0 = \dim(\Pi_\varkappa/R(V)), \quad n_1 = \dim(\Pi_\varkappa/D(V)) .$$

In other words, the defect number n_0 (n_1) is the dimension of the orthogonal or H-orthogonal complement of R(V) (D(V), respectively).

If only one of the numbers n_0 or n_1 is equal to zero, then, obviously, the isometric operator V is maximal, i.e., it has no proper isometric extension in $\Pi_\varkappa$. In this case either V (if $n_1 = 0$) or V^{-1} (if $n_0 = 0$) is a semi-unitary operator. If $n_0 = n_1 = 0$ then V is a unitary operator and hence also maximal isometric.

If both defect numbers are different from zero, the operator V has isometric extensions. The description of these extensions offers no difficulties if the scalar product does not degenerate on D(V) (and R(V)). In this case D(V) and R(V) are two spaces of type $\Pi_{\varkappa_1}$

($0 \leq \varkappa_1 \leq \varkappa$), so that their orthogonal complements

$$D = \Pi_\varkappa \ominus D(V) \quad \text{and} \quad R = \Pi_\varkappa \ominus R(V)$$

are spaces of type $\Pi_{\varkappa-\varkappa_1}$.

In each of these we can choose subspaces (F and G respectively) of equal dimension, so that an isometric mapping $V_1F = G$ exists.

Then the operator

$$\tilde{V}f = \begin{cases} Vf & (f \in D(V)) \\ V_1f & (f \in F) \end{cases}$$

is an isometric extension of V.

It is clear that the construction described here exhausts all the possible isometric extensions of V in $\Pi_\varkappa$.

In particular, all the maximal isometric extensions are obtained in this way; the existence of these is obvious, since either D is isometric with a part of R, or R is isometric with a part of D.

<u>9.2.</u> If the scalar product degenerates on D(V) and D_0 is the isotropic subspace of D(V), then by Theorem 3.3 there is a decomposition

$$D(V) = D' \oplus D_0 , \tag{9.1}$$

which is such that the scalar product does not degenerate on D'. If now we put $R' = VD'$ and $R_0 = VD_0$, then it is evident that $R(V) = R' \oplus R_0$, where R_0 is the isotropic subspace of R(V) and the scalar product does not degenerate on R'.

Consider the restriction V' of the operator V:

$$D(V') = D', \quad R(V') = R', \quad V'f = Vf \quad (f \in D') .$$

Since the scalar product does not degenerate on D' (and R'), it is easy to describe all the isometric extensions of the operator V' in the way as this was done above. If we choose from these extensions those for which D_0 goes into R_0, then we get all the isometric extensions of V. We shall show that among these isometric extensions V are maximal ones, i.e., such that V or V^{-1} is semi-unitary.

Indeed, we choose subspaces E_0 (S_0) in $\Pi_\varkappa \ominus D'$ ($\Pi_\varkappa \ominus R'$ respectively), which are skewly linked with D_0 (with R_0) (see Lemma 3.1). Then the scalar product does not degenerate on $D_0 \dotplus E_0$ and $R_0 \dotplus S_0$. If

$d_1, \dots, d_m$ is a basis of D_0, then $r_j = Vd_j$, $j = 1, \dots, m$, is a basis of R_0. We choose bases $e_1, \dots, e_m$ and $s_1, \dots, s_m$ of E_0 and S_0 such that

$$(d_j, e_k) = (r_j, s_k) = \delta_{jk}, \quad j,k = 1, \dots, m . \tag{9.2}$$

No vector e_0 of E_0 different from zero is in $D(V)$. In fact, if $e_0 \in D(V)$, then by (9.1) $e_0 = d' + d_0$, where $d' \in D'$ and $d_0 \in D_0$. Then for any e_j we would have

$$(d_0, e_j) = (e_0 - d', e_j) = (e_0, e_j) - (d', e_j) = 0 \quad (j = 1, \dots, m) ,$$

since the scalar product on E_0 is identically null and $d' \in D' \perp E_0$. Thus $d_0 \perp E_0$ and $d_0 = \theta$ as D_0 and E_0 are skewly linked. Hence $e_0 = d' = \theta$, which contradicts the assumption.

Now we extend the operator V, defining it on E_0 by setting $Ve_j = s_j$ $(j = 1, \dots, m)$. This will leave V isometric, because of (9.2).

Since the scalar product does not degenerate on the subspaces

$$D(V) \dotplus E_0 = D' \oplus (D_0 \dotplus D_0), \quad R(V) \dotplus S_0 = R' \oplus (R_0 \dotplus S_0),$$

the existence of a maximal extension becomes obvious, as in 9.1.

Just as in Hilbert space, an isometric invertible operator is unitary if and only if its defect index is (0,0); the equality of the defect numbers n_0 and n_1 is necessary and sufficient for the isometric operator V in $\Pi_\varkappa$ to have a unitary extension in $\Pi_\varkappa$.

9.3. Let A now be a closed Hermitian operator in $\Pi_\varkappa$, $\overline{D(A)} = \Pi_\varkappa$, and let $\Pi_+ \subset D(A)$ be chosen as in 6.2. The defect index of A is defined as the pair (n_+, n_-), where the defect numbers n_+ and n_- are given by Theorem 6.1 and the following Note 1. As was mentioned in the proof of Theorem 6.1, the defect number n_+ equals $\dim(R(A_0 - \zeta I))^\perp =$ $= \dim(R(A_{22} - \zeta I))^\perp$ if $\operatorname{Im} \zeta > 0$ and ζ is not an eigenvalue of A. Therefore the defect index (n_+, n_-) of A coincides with the defect index of the closed H-Hermitian operator A_{22} in the Hilbert space Π_-.

The H-Hermitian operator A_{22} in Π_- has always selfadjoint extensions, possibly in a larger Hilbert space $\widetilde{\Pi}_- \supset \Pi_-$. The H-scalar product on $\widetilde{\Pi}_-$ we denote again by $[x,y]$ $(x, y \in \Pi_-)$. Consider a selfadjoint extension $\widetilde{A}_{22}$ of A_{22} in $\widetilde{\Pi}_-$. On the direct sum $\widetilde{\Pi}_\varkappa = \Pi_+ \dotplus \widetilde{\Pi}_-$ we define an H-scalar product

$$[\widetilde{x}, \widetilde{y}] = [x_+, y_+] + [\widetilde{x}_-, \widetilde{y}_-] \quad (\widetilde{x} = x_+ + \widetilde{x}_-, \ \widetilde{y} = y_+ + \widetilde{y}_-)$$

and an indefinite scalar product

$$(\tilde{x},\tilde{y}) = (x_+,y_+) - [\tilde{x}_-,\tilde{y}_-] \ .$$

Evidently, $\tilde{\Pi}_\varkappa$ satisfies axioms I - VI of 1.1 with respect to the scalar product $(\tilde{x},\tilde{y})$.

Now consider A_{21} as an operator from Π_+ into $\tilde{\Pi}_-$ and denote its adjoint (from $\tilde{\Pi}_-$ into Π_+) by $\tilde{A}_{21}^{[*]}$. Then

$$\tilde{A} = \begin{bmatrix} A_{11} & -\tilde{A}_{21}^{[*]} \\ A_{21} & \tilde{A}_{22} \end{bmatrix} \tag{9.3}$$

is a selfadjoint extension of A in the space $\tilde{\Pi}_\varkappa \supset \Pi_\varkappa$.

Extensions of a Hermitian operator A which act in a space $\tilde{\Pi}_\varkappa$ ($\supset \Pi_\varkappa$) with the same index $\varkappa$ as the originally given space $\Pi_\varkappa$ are called <u>regular</u>.

If $\tilde{A}_{22}$ is an arbitrary H-Hermitian extension of A_{22} in $\tilde{\Pi}_-$ ($\supset \Pi_-$), then, evidently, the operator $\tilde{A}$ given by (9.3) is a regular Hermitian extension of A in $\tilde{\Pi}_\varkappa$ ($\supset \Pi_\varkappa$), and it follows as above that in this way we get all the regular Hermitian extensions of A. Here $\tilde{A}$ is selfadjoint if and only if $\tilde{A}_{22}$ is H-selfadjoint. Now the following theorem is an immediate consequence of the above considerations and the corresponding results in Hilbert space theory.

<u>Theorem 9.1.</u> Let A be a Hermitian operator in $\Pi_\varkappa$. Then:
(i) A has regular selfadjoint extensions. Among these there exist selfadjoint extensions in $\Pi_\varkappa$ if and only if $n_+ = n_-$.

(ii) A is maximal (that is, does not have proper Hermitian extensions in $\Pi_\varkappa$) if and only if A is closed and $\min(n_+,n_-) = 0$.

(iii) A is selfadjoint if and only if A is closed and $n_+ = n_- = 0$.

(iv) A has maximal extensions in $\Pi_\varkappa$.

The representation (9.3) of an arbitrary regular selfadjoint extension $\tilde{A}$ of A shows that the difference of $\tilde{A}$ and the H-selfadjoint operator $A_{11} \oplus \tilde{A}_{22}$ is of norm $\leqq \|A_{21}\|$. Hence from [17a], Theorem 9.4, we have

<u>Theorem 9.2.</u> The spectrum $\sigma(\tilde{A})$ of an arbitrary regular selfadjoint extension $\tilde{A}$ of A lies in the strip $\Sigma_A = \{z \mid |\mathrm{Im}\, z| \leqq \|A_{21}\|\}$. If $z \notin \Sigma_A$, then

$$\|(\tilde{A} - zI)^{-1}\| \leqq (|\mathrm{Im}\, z| - \|A_{21}\|)^{-1} \ .$$

It can be shown that for points ζ outside the strip Σ_A the defect subspaces $(R(A - \zeta I))^{\perp}$ are positive. For this and further results see [24d], and also [9].

__Note.__ If V is an isometric operator in Π_κ, also the notion of a __regular isometric extension__ of V, acting in some possibly larger space $\widetilde{\Pi}_\kappa \supset \Pi_\kappa$, can be introduced. With the techniques of 9.1 - 2 an analogue of Theorem 9.1 can be proved. Then Theorem 9.1 follows from these results by Cayley transformation.

C H A P T E R   III

I N V A R I A N T   S U B S P A C E S   O F   L I N E A R   O P E R A T O R S   I N Π_κ

§ 10. Preliminary observations

10.1. The main specific property of unitary and, more generally, of expansive operators in a space Π_κ is the existence of a κ-dimensional nonnegative invariant subspace (Theorem 11.2 below). In this section we prepare the proof of this theorem.

Let us fix some decomposition (1.1):

$$\Pi_\kappa = \Pi_+ \oplus \Pi_- \tag{1.1}$$

and the corresponding H-scalar product (2.1) and norm (2.3) in Π_κ. Consider a nonnegative subspace L of Π_κ. Then, for arbitrary $x \in L$, $x = x_+ + x_-$, $x_+ \in \Pi_+$, $x_- \in \Pi_-$, we have $(x,x) = [x_+,x_+] - [x_-,x_-] \geqq 0$, that is,

$$\|x_-\|^2 \leqq \|x_+\|^2 \tag{10.1}$$

and

$$\|x_+\|^2 \leqq \|x_+\|^2 + \|x_-\|^2 = \|x\|^2 \leqq 2\|x_+\|^2 .$$

These inequalities show that on L the mapping $x \to x_+$ is one-to-one and continuous together with its inverse.[1] Denoting its range by L_+ ($\subset \Pi_+$) and the inverse mapping by F, we have $Fx_+ = x$ $(x_+ \in L_+)$ and

$$x_- = x - x_+ = (F - I)x_+ \text{ if } x \in L .$$

Define on L_+ the operator $K = F - I$. It follows that $x_- = Kx_+$ and $x = x_+ + Kx_+$ $(x \in L)$. Finally, (10.1) gives $\|K\| \leqq 1$. Thus we have proved the first part of the following lemma (the easy proof of the second part will be left to the reader).

[1]) This fact was actually established already in the proof of Lemma 1.1. However the method used here extends to nonpositive subspaces and also to general spaces with a regular indefinite metric.

Lemma 10.1. To an arbitrary nonnegative subspace $L \subset \Pi_\varkappa$ there correspond a subspace $L_+ \subset \Pi_+$ and an operator K from L_+ into Π_-, $\|K\| \leq 1$, such that the decompositon (1.1) of $x \in L$ is

$$x = x_+ + Kx_+ \quad (x_+ \in L_+) \ . \tag{10.2}$$

Conversely, given L_+ and K with these properties, the set of elements x in (10.2) is a nonnegative subspace of $\Pi_\varkappa$.

Formally, the definition of K in Lemma 10.1 includes its domain L_+. Then the correspondence between nonnegative subspaces of $\Pi_\varkappa$ and linear operators K from a subspace $D(K) \subset \Pi_+$ into Π_-, $\|K\| \leq 1$, is one-to-one. The operator K is called the angle operator of the nonnegative subspace L. It is easy to see that the subspace L is positive if and only if for its angle operator K we have $\|K\| < 1$.

If the nonnegative subspace $\tilde{L} \subset \Pi_\varkappa$ is an extension of L, then, evidently, the angle operator $\tilde{K}$ of $\tilde{L}$ is an extension of K; that is, $D(\tilde{K}) = \tilde{L}_+ \supset D(K) = L_+$, and $\tilde{K}x_+ = Kx_+$ if $x \in L_+$. Therefore, a nonnegative subspace $L \subset \Pi_\varkappa$ has a nonnegative proper extension $\tilde{L}$ if and only if $L_+ \neq \Pi_+$. In this case a desired extension $\tilde{L}$ is defined, e.g., by the angle operator

$$\tilde{K}x_+ = \begin{cases} Kx_+, & x_+ \in L_+ \\ \theta\ , & x_+ \in \Pi_+ \ominus L_+ \end{cases} .$$

This observation and the fact that $\dim \Pi_+ = \varkappa$ prove the following lemma.

Lemma 10.2. An arbitrary nonnegative (positive) subspace $L \subset \Pi_+$ is contained in a $\varkappa$-dimensional nonnegative (positive resp.) subspace of $\Pi_\varkappa$.

The $\varkappa$-dimensional nonnegative subspaces of $\Pi_\varkappa$ are the maximal nonnegative subspaces; that is, they are not properly contained in another nonnegative subspace (Lemma 1.2). The maximal nonpositive subspaces, however, cannot be characterized simply by their dimension. But, repeating the above arguments, it follows that the statement of Lemma 10.1 remains valid after replacing "nonnegative" by "nonpositive" and exchanging the indices + and −.

Similarly to the above considerations, with each maximal nonpositive subspace L' $(\subset \Pi_\varkappa)$ there is associated its angular operator K' such that $L' = \{x \mid x = K'x_- + x_-\}$, $x_- \in \Pi_-$. Here K' maps Π_- into Π_+, and $\|K'\| \leq 1$. Moreover, it is easy to see that the orthogonal complement $L^\perp$ of a $\varkappa$-dimensional nonnegative subspace L is a maximal nonpositive subspace, and if K is the angle operator of L, then its adjoint operator K^*

is the angle operator of $L^{\perp}$; that is, the elements x of $L^{\perp}$ have a representation

$$x = K^* x_- + x_-, \quad x_- \in \Pi_- . \qquad (10.3)$$

If the $\varkappa$-dimensional subspace L in $\Pi_\varkappa$ with angle operator K is positive, then the orthogonal complement $L^{\perp}$ is negative. Indeed, with the representation (10.3) we have

$$(x,x) = \|K^* x_-\|^2 - \|x_-\|^2 \leqq (\|K^*\|^2 - 1)\|x_-\|^2 < 0 \text{ if } x_- \neq \theta .$$

10.2. Now let T be a continuous operator in $\Pi_\varkappa$ with $D(T) = \Pi_\varkappa$. Suppose T maps some $\varkappa$-dimensional nonnegative subspace L into a nonnegative subspace $\hat{L}$. By Lemma 10.2, we can suppose that $\hat{L}$ is again $\varkappa$-dimensional. We write the representation (10.2) of L in the form

$$x = \begin{bmatrix} x_+ \\ Kx_+ \end{bmatrix} \qquad (x_+ \in \Pi_+)$$

and use for T the corresponding matrix representation (compare § 6)

$$T = \begin{bmatrix} T_{11} & T_{12} \\ T_{21} & T_{22} \end{bmatrix} . \qquad (10.4)$$

Then

$$Tx = \begin{bmatrix} T_{11}x_+ + T_{12}Kx_+ \\ T_{21}x_+ + T_{22}Kx_+ \end{bmatrix} ,$$

and if $\hat{K}$ denotes the angle operator of $\hat{L}$, then $Tx \in \hat{L}$ implies

$$Tx = \begin{bmatrix} y_+ \\ \hat{K}y_+ \end{bmatrix} \qquad \text{for some } y_+ \in \Pi_+ .$$

Together with (10.4) this gives

$$T_{21}x_+ + T_{22}Kx_+ = \hat{K}(T_{11}x_+ + T_{12}Kx_+) \qquad \text{for all } x_+ \in \Pi_+ ,$$

or

$$T_{21} + T_{22}K = \hat{K}(T_{11} + T_{12}K) . \qquad (10.5)$$

This reasoning can be reversed, and we get

Lemma 10.3. The bounded linear operator T in $\Pi_\varkappa$ maps the $\varkappa$-dimensional nonnegative subspace L into the $\varkappa$-dimensional nonnegative subspace $\hat{L}$ if and only if (10.5) holds (here K and $\hat{K}$ denote the angle operators of L and $\hat{L}$, respectively).

Note 1. If L has the property that $Tx = \theta$ and $x \in L$ imply $x = \theta$, then the operator $T_{11} + T_{12}K$ does not have the eigenvalue zero; that is, it has a bounded inverse in Π_+ and (10.5) can be written as

$$\hat{K} = (T_{21} + T_{22}K)(T_{11} + T_{12}K)^{-1} . \tag{10.6}$$

Note 2. In the same way, T maps the maximal nonpositive subspace L' with angle operator K' into the maximal nonpositive subspace $\hat{L}'$ with angle operator $\hat{K}'$ if and only if

$$T_{11}K' + T_{12} = \hat{K}'(T_{21}K + T_{22}) .$$

In the following the set of all contractions from Π_+ into Π_- (that is, the set of all angle operators of $\varkappa$-dimensional nonnegative subspaces) will be denoted by $\mathcal{K}$.

Note 3. In the general case of a space Π with regular indefinite metric the angle operator was first introduced and applied by R.S. Phillips [34a] and, independently, by Ju.P. Ginzburg [16b].

§ 11. Invariant semidefinite subspaces of expansive, contractive and dissipative operators in $\Pi_\varkappa$ [1])

11.1. Before stating the main theorems we prove some lemmas which are perhaps of independent interest. We recall (see Theorem 6.6) that for each plus-operator T there exists a number $\mu \geqq 0$ such that

$$(Tx,Tx) \geqq \mu(x,x) \quad (x \in \Pi_\varkappa) ; \tag{11.1}$$

one can choose, e.g., $\mu = \mu(T) = \inf\{(Tx,Tx) \mid (x,x) = 1\}$.

Lemma 11.1. Let T be a plus-operator in $\Pi_\varkappa$ and suppose the number $\mu \geqq 0$ satisfies (11.1). Then each point $\zeta, |\zeta| > \sqrt{\mu}$, belongs to $\rho(T) \cup \sigma_d(T)$.

Proof. With the notation of 6.2 we have from (11.1)

$$\|T_{22}x_-\|^2 \leqq \mu\|x_-\|^2 + \|T_{12}x_-\|^2 \quad (x_- \in \Pi_-) ,$$

or

$$T_{22}^*T_{22} \leqq \mu I_- + T_{12}^*T_{12} = R^2 .$$

[1])The authors express their sincere thanks to T.Ja. Azizov for thorough consideration of § 11 and his many valuable remarks

Here I_- is the unit operator and R the nonnegative operator $R = (\mu I_- + T_{12}^* T_{12})^{1/2}$ in the H-space Π_-.

Now $\|T_{22}x_-\|^2 \leqq \|Rx_-\|^2$ $(x \in \Pi_-)$ implies $T_{22} = SR$, $\|S\| \leqq 1$.

The operator T_{12} is finite-dimensional. Then the same is true for the nonnegative operator $T_{12}^* T_{12}$; that is, it has the form $\sum_{j=1}^{r} \alpha_j P_j$ with positive numbers α_j and one-dimensional orthoprojectors P_j in Π_-, $j = 1,\dots,r$; $r \leqq \varkappa$. Therefore

$$R^2 = \mu I_- + \sum_{j=1}^{r} \alpha_j P_j = \mu(I_- - \sum_{j=1}^{r} P_j) + \sum_{j=1}^{r} (\alpha_j + \mu) P_j$$

and

$$R = \sqrt{\mu}(I_- - \sum_{j=1}^{r} P_j) + \sum_{j=1}^{r} (\alpha_j - \mu)^{1/2} P_j = \sqrt{\mu}\, I_- + \sum_{j=1}^{r} (\sqrt{\alpha_j - \mu} - \sqrt{\mu}) P_j .$$

It follows that $T_{22} = SF = \sqrt{\mu}\, S + R_0$, where $\|S\| \leqq 1$ and $\dim R(R_0) \leqq r \leqq \varkappa$.

If P_- denotes the H-orthogonal projector from $\Pi_\varkappa$ onto Π_-, we have $T_{22}P_- = \sqrt{\mu}\, SP_- + R_0 P_-$; here again $\|SP_-\| \leqq 1$, $\dim R(R_0 P_-) \leqq r$. The difference $T - T_{22}P_-$ is a finite-dimensional operator, and the statement of the lemma follows from well-known results of perturbation theory (see [17a]).

Lemma 11.2. Let T be a uniformly expansive operator in $\Pi_\varkappa$, that is,

$$(Tx,Tx) \geqq (x,x) + \delta^2 \|x\|^2 \quad (x \in \Pi_\varkappa) \tag{11.2}$$

for some H-norm $\|\cdot\|$ on $\Pi_\varkappa$; here we can suppose $0 < \delta < 1$. Then the ring $\Gamma : \sqrt{1 - \delta^2} < |\zeta| < \sqrt{1 + \delta^2}$ belongs to $\rho(T)$.

Proof.[1]) The condition (11.2) implies

$$(Tx,Tx) \geqq (x,x) + \delta^2 \|x\|^2 \geqq (1 - \delta^2)(x,x) \quad (x \in \Pi_\varkappa) . \tag{11.3}$$

According to Lemma 11.1, outside the disc $|\zeta| \leqq \sqrt{1 - \delta^2}$ the spectrum $\sigma(T)$ consists only of isolated eigenvalues.

[1])This proof was communicated to us by T.Ja. Azizov. It is shorter than our original proof.

On the other hand, if $Tx = \lambda x$, $x \neq \theta$, then from (11.2) we have

$$\delta^2 \|x\|^2 \leq (|\lambda|^2 - 1)(x,x) \leq ||\lambda|^2 - 1| \|x\|^2 ;$$

that is, $||\lambda|^2 - 1| \geq \delta^2$ or $\lambda \notin \Gamma$. The lemma is proved.

<u>Lemma 11.3</u> (cf. Lemma 6.3). If T is an expansive or uniformly expansive operator in $\Pi_\varkappa$, so is T^*.[1]

<u>Proof.</u> Let T be a uniformly expansive operator, i.e.

$$(Tx,Tx) \geq (x,x) + \delta^2 \|x\|^2 \qquad (x \in \Pi_\varkappa)$$

or, equivalently,

$$((T^*T - I)x,x) \geq \delta^2 \|x\|^2 \qquad (x \in \Pi_\varkappa) \tag{11.4}$$

for some $\delta > 0$. According to Lemma 3.5, $\zeta = 1$ belongs to $\rho(T)$. Consider the continuous operator

$$A = (T + I)(T - I)^{-1} , \qquad D(A) = \Pi_\varkappa .$$

Then $1 \in \rho(A)$, $T = (A + I)(A - I)^{-1}$, and (11.4) can be written as follows:

$$([(A^* - I)^{-1}(A^* + I)(A + I)(A - I)^{-1} - I]x,x) \geq \delta^2 \|x\|^2 \qquad (x \in \Pi_\varkappa)$$

or

$$([(A^* + I)(A + I) - (A^* - I)(A - I)](A - I)^{-1}x, (A - I)^{-1}x) \geq \delta^2 \|x\|^2 \qquad (x \in \Pi_\varkappa) .$$

Using the fact that $(A - I)\Pi_\varkappa = \Pi_\varkappa$ and $\|(A - I)y\| \geq C\|y\|$ $(y \in \Pi_\varkappa)$ for some $C > 0$, we get

$$2((A + A^*)y,y) \geq \delta^2 C^2 \|y\|^2 \qquad (y \in \Pi_\varkappa) . \tag{11.5}$$

This relation is symmetric with respect to A and A*. The reasoning leading from (11.4) to (11.5) can be reversed, interchanging the roles of A and A*. Then we obtain

---

[1]) For expansive operators this statement was first proved by Ju.P. Ginzburg [16a]. He obtained it as a consequence of his general criterion for the fact that in a space Π with regular indefinite metric the adjoint T^* of an expansive operator T is also expansive (comp. also with the remarks to Lemma 11.9 in the "Historical Notes" to § 11).

$$((TT^* - I)x,x) \geqq \delta_1^2 \|x\|^2 \qquad (x \in \Pi_\varkappa)$$

with some $\delta_1 > 0$. Thus T^* is uniformly expansive simultaneously with T.

Now let T be an arbitrary expansive operator. If $0 < \delta < 1$ and S_δ is given by (6.12), the operator TS_δ is uniformly expansive (see 6.5). Then $S_\delta^* T^*$ is expansive by the first part of the proof; that is,

$$(S_\delta^* T^* x,\ S_\delta^* T^* x) \geqq (x,x) \qquad (x \in \Pi_\varkappa)\ .$$

Letting $\delta \downarrow 0$, we obtain $(T^*x,T^*x) \geqq (x,x)$ $(x \in \Pi_\varkappa)$, and the lemma is proved.

<u>Lemma 11.3'.</u> Let T be a strict plus-operator. If for some $\mu' > 0$

$$(Tx,Tx) \geqq \mu'(x,x) \qquad (x \in \Pi_\varkappa)\ ,$$

then also

$$(T^*x,T^*x) \geqq \mu'(x,x) \qquad (x \in \Pi_\varkappa)\ ,$$

and $\mu(T) = \mu(T^*)$.

<u>Proof.</u> The first inequality means that $T_1 = (\mu')^{-1/2}T$ is an expansive operator. Then also the operator $T_1^* = (\mu')^{-1/2}T^*$ is expansive, and the second inequality follows. Moreover, choosing $\mu' = \mu(T)$, it follows that $\mu(T) \leqq \mu(T^*)$, and exchanging T and T^* we get the last statement of the lemma,

<u>Corollary 1.</u> Let T be a strict plus-operator. Then also T^*, T^*T and TT^* are strict plus-operators, and $\mu_\pm(T) = \mu_\pm(T^*)$.

The first statements follow immediately from Lemma 11.3'. Choosing there $\mu' = \mu_\pm(T)$, we get $\mu_+(T^*) \geqq \mu_+(T)$ and $\mu_-(T^*) \leqq \mu_-(T)$, and the last conclusion follows after exchanging T and T^*.

The example in footnote on p. 45 shows that for a plus-operator T the operator T^* need not be a plus-operator.

<u>Corollary 2.</u> Let T be a plus-operator and $\mu \in [\mu_-(T),\mu_+(T)]$. Then $A = T^*T$ satisfies

$$((A - \mu I)x,x) \geqq 0 \qquad (x \in \Pi_\varkappa)\ .$$

If, moreover, T is a strict plus-operator, then

$$(A(A - \mu I)x,x) \geqq 0 \qquad (x \in \Pi_\varkappa)\ .$$

Indeed, the first statement was already established in (6.15). The second relation follows from

$$(T^*Tx, T^*Tx) \geqq \mu(Tx, Tx) = \mu(T^*Tx, x) \qquad (x \in \Pi_\varkappa) .$$

The relations in Corollary 2 have interesting consequences about spectral properties of T^*T. These will be formulated in 12.6.

11.2. Theorem 11.1. Let T be a continuous operator in $\Pi_\varkappa$ with the following properties:

a) $(Tx, Tx) > (x, x) \quad (x \in \Pi_\varkappa,\ x \neq \theta)$;

b) $\sigma(T) \cap \{\zeta \mid |\zeta| = 1\} = \emptyset$.

Then $\Pi_\varkappa$ is the direct sum $\mathfrak{I}_+ \dotplus \mathfrak{I}_-$ of two subspaces $\mathfrak{I}_+$ and $\mathfrak{I}_-$, invariant under T and such that: 1) $\mathfrak{I}_+$ is a $\varkappa$-dimensional positive subspace; 2) $\mathfrak{I}_-$ is a maximal negative subspace; 3) $|\sigma(T|\mathfrak{I}_+)| > 1$, $|\sigma(T|\mathfrak{I}_-)| < 1$.

The subspace $\mathfrak{I}_+$ ($\mathfrak{I}_-$) is the unique maximal nonnegative (nonpositive respectively) subspace which is invariant under the operator T.

Proof. Let $|\zeta| = 1$. Then according to b) the operator $(T - \zeta I)^{-1}$ exists, and from a) we have for $y \neq \theta$

$$(T(T - \zeta I)^{-1}y,\ T(T - \zeta I)^{-1}y) > ((T - \zeta I)^{-1}y,\ (T - \zeta I)^{-1}y) ,$$

which can be written as

$$(y,y) + \zeta((T - \zeta I)^{-1}y, y) + \bar{\zeta}(y, (T - \zeta I)^{-1}y) > 0 . \tag{11.6}$$

We introduce the Riesz projector

$$P = -\frac{1}{2\pi i}\int_{|\zeta|=1} (T - \zeta I)^{-1}\, d\zeta .$$

Putting $\zeta = e^{i\varphi}$, in (11.6) multiplying both sides by $d\varphi/2\pi$ and integrating from 0 to 2π, we get

$$(y,y) > (y, Py) + (Py, y) = 2\,\mathrm{Re}(Py, y) . \tag{11.7}$$

The subspaces $\mathfrak{I}_- = P\Pi_\varkappa$ and $\mathfrak{I}_+ = (I - P)\Pi_\varkappa$ are invariant under T. If $y \in \mathfrak{I}_-$ (that is, $Py = y$) and $y \neq \theta$, then (11.7) implies $(y,y) < 0$; if $y \in \mathfrak{I}_+$ or $Py = \theta$, $y \neq \theta$, then (11.7) implies $(y,y) > 0$.

As the direct sum of $\mathfrak{I}_+$ and $\mathfrak{I}_-$ is $\Pi_\varkappa$, it is easy to see that $\mathfrak{I}_+$ ($\mathfrak{I}_-$) is a $\varkappa$-dimensional positive (maximal negative, respectively) subspace. Indeed, if e.g. $\mathfrak{I}_- \subset L_-$, where L_- is a negative subspace and $L_- \neq \mathfrak{I}_-$, then for $x \in L_- \setminus \mathfrak{J}_-$ we have $x = x_1 + x_2$, $x_1 \in \mathfrak{J}_+$, $x_2 \in \mathfrak{J}_- \subset L$. Hence $x_1 = x - x_2 \in L_-$. On the other hand, $\mathfrak{J}_+ \cap L_- = \{\theta\}$, which implies $x_1 = \theta$ and $x = x_2 \in \mathfrak{J}_-$, a contradiction.

The assertion 3) follows from well-known properties of Riesz projectors [11]. The simple proof of the last statement of Theorem 11.1 will be omitted here because according to Theorem 11.4 below this statement holds even under more general conditions (if only a) is satisfied).

From Lemma 11.2 and Theorem 11.1 we have the

Corollary. If T is uniformly expansive in $\Pi_\varkappa$, then the conclusions of Theorem 3.1 hold. Moreover, if δ is given by (11.2), then

$$|\sigma(T \mid \mathfrak{J}_+)| \geqq \sqrt{1+\delta^2} \quad \text{and} \quad |\sigma(T \mid \mathfrak{J}_-)| \leqq \sqrt{1-\delta^2} \; .$$

Note 1. By the same considerations as in the proof of Theorem 11.1 a Theorem 11.1' can be proved where condition a) is replaced by

a') $(Tx, Tx) < (x,x) \quad (x \in \Pi_\varkappa,\ x \neq \theta)$,

condition b) and the statements 1), 2) remain unchanged and 3) is replaced by

3') $|\sigma(T \mid \mathfrak{J}_+)| < 1, \quad \sigma|(T \mid \mathfrak{J}_-)| > 1$.

Note 2. In the proof of Theorem 11.1 we did not make use of the fact that $\varkappa < \infty$. Therefore Theorem 11.1 remains valid in spaces Π with a regular indefinite metric (see 1.1) if 1) is replaced by

1') $\mathfrak{J}_+$ is a maximal positive subspace.

In this form the theorem has important applications in stability theory [11].

From the point of view of operators in general spaces Π the conditions a) and a') can be transformed into each other, if we replace the scalar product (x,y) by $(x,y)' = -(x,y) \quad (x,y \in \Pi)$.

Corollary. For a selfadjoint operator A ($D(A) = \Pi_\varkappa$) the following statements are equivalent:

I) A is uniformly expansive.

II) A satisfies the conditions a) and b) of Theorem 11.1.

III) The space $\Pi_\varkappa$ admits a decomposition $\Pi_\varkappa = \mathfrak{J}_+ \oplus \mathfrak{J}_-$ with a positive subspace $\mathfrak{J}_+$ and a negative subspace $\mathfrak{J}_-$, both invariant under A

and with the properties $|\sigma(A \mid \mathfrak{J}_-)| < 1$ and $|\sigma(A \mid \mathfrak{J}_+)| > 1$.
If these conditions I) - III) are satisfied, then $\mathfrak{J}_+$ ($\mathfrak{J}_-$) is the unique maximal nonnegative (nonpositive, respectively) subspace which is invariant under A. Moreover, $\mu_\pm(A) = \alpha_\pm$, where

$$\alpha_- = \max|\sigma(A \mid \mathfrak{J}_-)| \; (< 1 <) \; \alpha_+ = \min|\sigma(A \mid \mathfrak{J}_+)| \; .$$

Indeed, as was already mentioned, I) implies II) according to Lemma 11.2. If $A = A^*$, the Riesz projector in the proof of Theorem 11.1 for $T = A$ is selfadjoint: $P = P^*$, hence orthogonal; that is, $R(P) \perp R(I-P)$. Therefore, in this case we have $\Pi_\varkappa = \mathfrak{J}_+ \oplus \mathfrak{J}_-$, and II) implies III).

If III) is satisfied, then A is the orthogonal sum $A_+ \oplus A_-$ of its restrictions $A_\pm = A \mid \mathfrak{J}_\pm$. Evidently, A_+ and A_- are selfadjoint operators in the Hilbert spaces $\mathfrak{J}_+$ (with scalar product (x,y)) and $\mathfrak{J}_-$ (with scalar product $-(x,y)$), respectively; therefore

$$(Ax,Ax) \geqq \alpha_+^2(x,x) \quad (x \in \mathfrak{J}_+), \qquad (Ax,Ax) \geqq \alpha_-^2(x,x) \quad (x \in \mathfrak{J}_-) ,$$

where $\alpha_\pm^2 \in \sigma(A_\pm^2)$, and in the first inequality the sign = holds for the eigenvector x corresponding to the eigenvalue α_+^2 of the finite-dimensional operator A_+^2, and in the second relation α_-^2 cannot be replaced by any smaller number. It follows that $\mu_+(A) \leqq \alpha_+^2$ and $\mu_-(A) \geqq \alpha_-^2$. On the other hand, for arbitrary $x = x_+ + x_-$ $(x_\pm \in \mathfrak{J}_\pm)$, from $\alpha_- < 1 < \alpha_+$ we have

$$(Ax,Ax) = (A_+x_+,A_+x_+) + (A_-x_-,A_-x_-) \geqq \alpha_+^2(x_+,x_+) + \alpha_-^2(x_-,x_-) \geqq$$
$$\geqq \alpha_\pm^2[(x_+,x_+) + (x_-,x_-)] = \alpha_\pm^2(x,x) \; ;$$

hence $\mu_+(A) \geqq \alpha_+^2$ and $\mu_-(A) \leqq \alpha_-^2$. Therefore III) implies the last statement of the corollary. Moreover,

$$\alpha_+^2(x_+,x_+) + \alpha_-^2(x_-,x_-) = (x,x) + (\alpha_+^2 - 1)(x_+,x_+) - (1 - \alpha_-^2)(x_-,x_-) \geqq$$
$$\geqq (x,x) + \delta\|x\|^2 ,$$

where $\delta = \min(\alpha_+^2 - 1, 1 - \alpha_-^2)$ and $\|x\|^2 = (x_+,x_+) - (x_-,x_-)$, and so III) also implies I). The corollary is proved.

This corollary describes the structure of selfadjoint uniformly expansive operators completely.

<u>11.3. Lemma 11.4.</u> Let A be a continuous operator in $\Pi_\varkappa$, $D(A) = \Pi_\varkappa$, L a $\varkappa$-dimensional nonnegative invariant subspace of A with angle operator $K \mid \Pi_+ \to \Pi_-$ and $A = \|A_{jk}\|^2_{j,k=1}$ the matrix representation of A. Then

$$\sigma(A \mid L) = \sigma(A_{11} + A_{12} K) .$$

Moreover, if A has a maximal nonpositive invariant subspace L' with angle operator $K' \mid \Pi_- \to \Pi_+$, then

$$\sigma(A \mid L') = \sigma(A_{21}K' + A_{22}), \quad \sigma_d(A \mid L') = \sigma_d(A_{21}K' + A_{22}) .$$

Proof. We shall prove only the second statement, the proof of the first being similar. The mapping S: $Sx = x_-$, defined for $x \in L'$, $x = K'x_- + x_-$, maps L' one-to-one and continuously onto Π_-. If $x \in L'$, then $Ax \in L'$ and $SAx = (A_{21}K' + A_{22})x_-$; that is, we have

$$A \mid L' = S^{-1}(A_{21}K' + A_{22})S .$$

In this case the spectra σ, σ_p and σ_d of the operators $A \mid L'$ and $A_{21}K' + A_{22}$ coincide, including the multiplicities of the eigenvalues.

In the following, two simple lemmas about invariant subspaces, eigen- and root vectors of expansive operators will play an essential role.

Lemma 11.5. If T is an expansive operator in $\Pi_\varkappa$ and L is a subspace of $\Pi_\varkappa$ such that TL = L and $(Tx,Tx) = (x,x)$ for all $x \in L$, then

$$T^*L = L \quad \text{and} \quad (T^*|L)(T|L) = (T|L)(T^*|L) = I_L . \tag{11.8}$$

Proof. As T is expansive, the operator $W = T^*T - I$ is nonnegative: $(Wx,x) \geq 0 \quad (x \in \Pi_\varkappa)$. For an arbitrary $g \in L$ we have $(Wg,g) = 0$. The last equality and the semidefiniteness of the form (Wx,x) imply $(Wg,x) = 0 \quad (x \in \Pi_\varkappa)$; that is, $Wg = \theta$ or

$$T^*T\, g = g \quad (g \in L) . \tag{11.9}$$

This implies (11.8).

Corollary 1. If L is a neutral subspace, invariant under the expansive operator T, then TL = L and (11.8) holds. In particular, if φ is a neutral eigenvector of T: $T\varphi = \lambda\varphi$, then $\lambda \neq 0$ and $T^*\varphi = \lambda^{-1}\varphi$.

We have only to show that for the neutral subspace L the inclusion $TL \subset L$ implies TL = L. As L is at most $\varkappa$-dimensional, this will follow if only $T|L$ does not have the eigenvalue zero. But this is a consequence of the following simple fact:

The kernel of an expansive operator T is a negative subspace.[1]

[1] This statement was first proved by Ju.P. Ginzburg.

Indeed, $T\varphi = \theta$ implies $(\varphi,\varphi) \leq 0$. Assume $(\varphi,\varphi) = 0$. Then $(W\varphi,\varphi) = 0$ and, as in the proof of Lemma 11.5, we find that $W\varphi = \theta$ or $T^*T\varphi = \varphi = \theta$.

Now let $\|x\|$, $x \in \Pi_\varkappa$, be an arbitrary H-norm on $\Pi_\varkappa$ (see (2.3)). It belongs to a decomposition $\Pi_\varkappa = \Pi_+ \oplus \Pi_-$. If T is an expansive operator in $\Pi_\varkappa$, for its corresponding operator norm $\|T\|$ we find that

$$\|x\|^2 \leq (Tx,Tx) \leq \|T\|^2\|x\|^2 \quad (x \in \Pi_+) ;$$

hence $\|T\| \geq 1$.

Corollary 2. If T is an expansive operator in $\Pi_\varkappa$, the eigenspaces $L_\lambda(T)$ corresponding to eigenvalues λ with the property $|\lambda| < \|T\|^{-1}$ are negative.

Proof. The definition of an expansive operator implies that the eigenspaces $L_\lambda(T)$, $|\lambda| < 1$, are nonpositive. Assume that for some λ, $0 < |\lambda| < \|T\|^{-1}$ and $\varphi \in L_\lambda(T)$ we have $(\varphi,\varphi) = 0$. Then, by Corollary 1, $T^*\varphi = \lambda^{-1}\varphi$, which is impossible because of $|\lambda|^{-1} > \|T\| = \|T^*\|$. If $0 \in \sigma_p(T)$, it was shown during the proof of Corollary 1 that $L_0(T)$ is negative, and the proof of the corollary is complete.

It is not hard to see that for a continuous linear operator T in $\Pi_\varkappa$ the root manifolds $S_{\lambda_1}(T)$ and $S_{\lambda_2}(T^*)$ are orthogonal if $\lambda_2 \neq \overline{\lambda_1}$. This can either be shown by arguments similar to those used in 7.2 for isometric operators, or it can be deduced from the corresponding fact in Hilbert space using relations (5.2). Hence from Lemma 11.5 we also have

Corollary 3. If T and L have the same properties as in Lemma 11.5, then L is orthogonal to each root manifold $S_\lambda(T)$ if only $\overline{\lambda}^{-1} \in \sigma(T|L)$.

Lemma 11.6. Let T be an expansive operator in $\Pi_\varkappa$ and $T\varphi = \varepsilon\varphi$, where $|\varepsilon| = 1$ and $\varphi \neq 0$. Then $T^*\varphi = \overline{\varepsilon}\varphi$, and φ is orthogonal to each root manifold $S_\lambda(T)$ with $\lambda \neq \varepsilon$. If, moreover, $(\varphi,\varphi) \neq 0$, then φ is not in the range of the operator $T - \varepsilon I$. Therefore, if the eigenspace $L_\varepsilon(T)$, $|\varepsilon| = 1$, is definite, it coincides with the corresponding root manifold: $S_\varepsilon(T) = L_\varepsilon(T)$.

Proof. The first two statements follow immediately from Lemma 11.5 and its Corollary 3. Assume φ is in the range of $T - \varepsilon I$: $T\varphi' = \varepsilon\varphi' + \varphi$. Scalar multiplication by φ implies $(T\varphi',\varphi) = \varepsilon(\varphi',\varphi) + (\varphi,\varphi)$. But we have

$$(T\varphi',\varphi) = (\varphi',T^*\varphi) = \varepsilon(\varphi,\varphi) ,$$

and hence $(\varphi,\varphi) = 0$. Therefore, if $L_\varepsilon(T)$ is definite, there cannot be a Jordan chain of length greater than one corresponding to ε, and the last statement follows.

11.4. In this section we study the semidefinite invariant subspaces of expansive operators. These results immediately apply to a strict plus-operator T, as in this case the operator $T' = \mu(T)^{-1/2}T$ is expansive. On the other hand, considering an arbitrary expansive operator T, we do not suppose that $\mu(T) = 1$ (it may happen that $\mu(T) > 1$). Therefore the consideration of expansive operators sometimes gives slightly more exact statements.

If the plus-operator T is not strict ($\mu(T) = 0$), then its range $R(T)$ is a nonnegative subspace, and an arbitrary $\varkappa$-dimensional nonnegative subspace, which contains $R(T)$, is invariant with respect to T. Moreover, it is easy to see that $R(T)$ contains all the root manifolds $S_\lambda(T)$ corresponding to nonzero eigenvalues λ of T. Therefore this case is not of special interest.

If T is an expansive operator in $\Pi_\varkappa$, we denote by $\mathfrak{S}_+(T)$ (by $\mathfrak{S}_-(T)$) the set of all $\varkappa$-dimensional nonnegative (maximal nonpositive) subspaces $\mathfrak{I}_+$ ($\mathfrak{I}_-$) which are invariant under T and have the property

$$|\sigma(T|\mathfrak{I}_+)| \geqq 0 \quad (|\sigma(T|\mathfrak{I}_-)| \leqq 1) . \tag{11.10}$$

The elements of $\mathfrak{S}_\pm(T^*)$ will be denoted by $\mathfrak{I}_\pm^*$.

Theorem 11.2. Let T be an expansive operator in $\Pi_\varkappa$. Then the sets $\mathfrak{S}_+(T)$ and $\mathfrak{S}_-(T)$ are not empty. An arbitrary $\mathfrak{I}_+ \in \mathfrak{S}_+(T)$ ($\mathfrak{I}_- \in \mathfrak{S}_-(T)$) contains all the root manifolds $S_\lambda(T)$ with $|\lambda| > 1$ ($|\lambda| < 1$). The mapping $\mathfrak{I}_+^* \to (\mathfrak{I}_+^*)^\perp$ ($\mathfrak{I}_-^* \to (\mathfrak{I}_-^*)^\perp$) defines a one-to-one correspondence between $\mathfrak{S}_+(T^*)$ and $\mathfrak{S}_-(T)$ ($\mathfrak{S}_-(T^*)$ and $\mathfrak{S}_+(T)$, respectively).

Proof. I. If the space $\Pi_\varkappa$ is nonseparable, we consider the decomposition $\Pi_\varkappa = \Pi_\varkappa' \oplus \Pi_0'$ with a separable subspace $\Pi_\varkappa'$, given at the end of 8.2. Then T' and T_0' are again expansive operators, and we have $\|T_0'\| \leqq 1$ with respect to the H-norm $\|x\| = -(x,x)^{1/2}$ on Π_0'. Evidently $\mathfrak{S}_+(T') \subset \mathfrak{S}_+(T)$ and $\mathfrak{S}_+(T'^*) \subset \mathfrak{S}_+(T^*)$. Therefore, in part II of the proof we can suppose that the space $\Pi_\varkappa$ is separable; in part III we shall return to the general case.

II. Let $\Pi_\varkappa$ be separable. As in the proof of Lemma 11.3 we consider the uniformly expansive operator $T(\delta) = TS_\delta$ $(0 < \delta < 1)$. The corollary of Theorem 11.1 implies that $\Pi_\varkappa$ decomposes as the direct sum of its subspaces $\mathfrak{I}_\pm(\delta)$, ($\mathfrak{I}_\pm(\delta) \in \mathfrak{S}_\pm(T(\delta))$), which are invariant under $T(\delta)$ and such that $\mathfrak{J}_+(\delta)$ is a $\varkappa$-dimensional positive and $\mathfrak{J}_-(\delta)$ a maximal negative subspace,

$$|\sigma(T(\delta) \mid \mathfrak{J}_+(\delta))| > 1, \quad |\sigma(T(\delta) \mid \mathfrak{J}_-(\delta))| < 1 .$$

Denote the angle operator of $\mathfrak{J}_+(\delta)$ with respect to some decomposition (1.1): $\Pi_\varkappa = \Pi_+ \oplus \Pi_-$ by $K(\delta)$. Then the invariance of the subspace $\mathfrak{J}_+(\delta)$ under $T(\delta)$ can be written as (see (10.5))

$$T_{21}(\delta) + T_{22}(\delta)\, K(\delta) = K(\delta)(T_{11}(\delta) + T_{12}(\delta)\, K(\delta)) . \tag{11.11}$$

Here $T_{jk}(\delta)$ are the entries in the matrix representation of $T(\delta)$ with respect to the chosen decomposition of $\Pi_\varkappa$:

$$T(\delta) = \begin{bmatrix} T_{11}(\delta) & T_{12}(\delta) \\ T_{21}(\delta) & T_{22}(\delta) \end{bmatrix} = \begin{bmatrix} \sqrt{1+\delta^2}\, T_{11} & \sqrt{1-\delta^2}\, T_{12} \\ \sqrt{1+\delta^2}\, T_{21} & \sqrt{1-\delta^2}\, T_{22} \end{bmatrix} \tag{11.12}$$

The set $\mathcal{K}$ of all the angle operators of $\varkappa$-dimensional nonnegative subspaces of $\Pi_\varkappa$ is compact in the weak operator topology. By the separability of $\Pi_\varkappa$ there exists a sequence of positive numbers $\delta_n \downarrow 0$ such that $K(\delta_n)$ converges in the weak operator topology to some $K_0 \in \mathcal{K}$ if $n \to \infty$.

Evidently the sequences $(T_{jk}(\delta_n))$ converge to T_{jk}, $j,k = 1,2$, in the uniform operator topology (see (11.12)), and, using the fact that $\|K(\delta_n)\| \leq 1$ $(n = 1,2,\ldots)$, it follows easily that $T_{12}(\delta_n)K(\delta_n)$ converges in the weak operator topology to $T_{12}K_0$. As all these operators act in the finite-dimensional space Π_+, this convergence holds also in the uniform operator topology. The same is true for the sequence $T_{11}(\delta_n) + T_{12}(\delta_n)K(\delta_n)$, which converges to $T_{11} + T_{12}K_0$. Then the right-hand side of (11.11) with δ_n instead of δ converges in the weak operator topology to $K_0(T_{11} + T_{12}K_0)$. In the same way, the left-hand side of (11.11) for $\delta = \delta_n$ converges in the weak operator topology to $T_{21} + T_{22}K_0$, and we get finally

$$T_{21} + T_{22}K_0 = K_0(T_{11} + T_{12}K_0) .$$

The operator K_0, $\|K_0\| \leq 1$, defines a $\varkappa$-dimensional nonnegative subspace $\mathfrak{J}_0$, which is invariant under T (see 10.2).

From $\Pi_\varkappa = \mathfrak{J}_+(\delta) \dotplus \mathfrak{J}_-(\delta)$ it follows that

$$\sigma(T(\delta)) = \sigma(T(\delta) \mid \mathfrak{J}_+(\delta)) \cup \sigma(T(\delta) \mid \mathfrak{J}_-(\delta)) ,$$

where the first set on the right-hand side lies outside the closed unit disc, and the second set is in the open unit disc. Consider the operator

$\hat{T}(\delta) = T_{11}(\delta) + T_{12}(\delta)K(\delta)$ $(\delta \geqq 0,\ K(0) = K_0)$ in $\Pi_\varkappa$. Then Lemma 11.4 implies

$$\sigma(T(\delta) \mid \mathfrak{J}_+(\delta)) = \sigma(\hat{T}(\delta)), \quad \sigma(T|\mathfrak{J}_0) = \sigma(\hat{T}(0)) . \qquad (11.13)$$

The operators $\hat{T}(\delta_n)$ and $\hat{T}(0)$ act in the finite-dimensional space Π_+, and $\hat{T}(0) = \lim_{n\to\infty} \hat{T}(\delta_n)$. Therefore each eigenvalue of $\hat{T}(0)$ is the limit of eigenvalues of the operators $\hat{T}(\delta_n)$. By the first relation in (11.13), these eigenvalues are outside of the closed unit disc, and the second relation in (11.13) implies finally that $|\sigma(T|\mathfrak{J}_0)| \geqq 1$. Thus we have shown that the set $\mathfrak{S}_+(T)$ is not empty.

With the expansive operator T the operator T* is also expansive, and the already proved part of the theorem can be applied to T*. Hence there exists a maximal nonnegative subspace $\mathfrak{J}_+^*$ which is invariant under T* and such that

$$|\sigma(T^*|\mathfrak{J}_+^*)| \geqq 1 . \qquad (11.14)$$

III. Now we return to the general case; that is, the space $\Pi_\varkappa$ need not be separable. We have already shown that there exist subspaces $\mathfrak{J}_+ \in \mathfrak{S}_+(T)$ and $\mathfrak{J}_+^* \in \mathfrak{S}_+(T^*)$. Consider the orthogonal complement $\mathfrak{J}_- = (\mathfrak{J}_+^*)^\perp$. It is a maximal nonpositive subspace invariant under T. We shall show that it has the property $|\sigma(T|\mathfrak{J}_-)| \leqq 1$.

Indeed, if $\lambda_0 \in \sigma(T|\mathfrak{J}_-)$, $|\lambda_0| > 1$, then from Lemma 11.1 we have $\lambda_0 \in \sigma_p(T|\mathfrak{J}_-)$; that is, there exists a vector $x_0 \in \mathfrak{J}_-$ such that $Tx_0 = \lambda_0 x_0$. It follows that $|\lambda_0|^2(x_0,x_0) = (Tx_0,Tx_0) \geqq (x_0,x_0)$, $(x_0,x_0) \leqq 0$, hence $(x_0,x_0) = 0$; that is, $x_0 \perp \mathfrak{J}_-$ or $x_0 \in \mathfrak{J}_- \cap \mathfrak{J}_+^*$. Now Corollary 1 of Lemma 11.5 implies $\lambda_0^{-1} \in \sigma(T^*|\mathfrak{J}_+^*)$, which is a contradiction to (11.14) because $|\lambda_0^{-1}| < 1$.

Thus the first statement of Theorem 11.2 is proved. We have also shown that $\mathfrak{J}_+^* \in \mathfrak{S}_+(T^*)$ implies $(\mathfrak{J}_+^*)^\perp \in \mathfrak{S}_-(T)$. Exchanging + and - and, finally, T and T*, the last statement of Theorem 11.2 follows.

IV. To prove the second statement of Theorem 11.2 we show e.g., that an arbitrary $\mathfrak{J}_- \in \mathfrak{S}_-(T)$ contains each root manifold $S_{\lambda_0}(T)$, where $\lambda_0 \in \sigma_p(T)$, $|\lambda_0| < 1$. As has already been shown, $\mathfrak{J}_- = (\mathfrak{J}_+^*)^\perp$ for some subspace $\mathfrak{J}_+^* \in \mathfrak{S}_+(T^*)$. Consider for arbitrary $x \in S_{\lambda_0}(T)$ and $y \in \mathfrak{J}_+^*$ the function

$$\lambda \to ((T - \lambda I)^{-1}x,y) = (x,(T^* - \bar{\lambda}I)^{-1}y) .$$

The left-hand side of this equality can be continued holomorphically to all $\lambda \neq \lambda_0$; the right-hand side is holomorphic in the open unit disc containing λ_0. Therefore Liouville's theorem implies

$((T - \lambda I)^{-1}x,y)_{,}=$ const. Multiplying this relation by λ and letting $|\lambda| \to \infty$, it follows that $(x,y) = 0$, which means $S_{\lambda_0}(T) \subset (\mathfrak{I}_+^*)^{\perp} = \mathfrak{I}_-$. In the same way it can be shown that each $\mathfrak{I}_+ \in \mathfrak{G}_+(T)$ contains all the root manifolds $S_{\lambda_0}(T)$ corresponding to $\lambda_0 \in \sigma_p(T)$, $|\lambda_0| > 1$.

Theorem 11.2 is proved.

Corollary 1. The linear span $E_+(T)$ of all root manifolds $S_\lambda(T)$, $|\lambda| > 1$, of the expansive operator T in $\Pi_\varkappa$ is a nonnegative subspace and hence of dimension $\leq \varkappa$. If $\dim E_+(T) = \varkappa$, then $E_+(T)$ is the unique element of $\mathfrak{G}_+(T)$. In this case each eigenvector φ of T: $T\varphi = \varepsilon\varphi$, with $|\varepsilon| = 1$, is negative: $(\varphi,\varphi) < 0$; moreover, also $\mathfrak{I}_- \in \mathfrak{G}_-(T)$ is uniquely determined, and $\Pi_\varkappa = \mathfrak{I}_+ \dotplus \mathfrak{I}_-$.

The first statement follows immediately from Theorem 11.2. Now let $\dim E_+(T) = \varkappa$. If $T\varphi = \varepsilon\varphi$, $\varphi \neq \theta$, $|\varepsilon| = 1$, then Lemma 11.6 implies $T^*\varphi = \bar{\varepsilon}\varphi$, and $\varphi \perp E_+^*(T)$. Therefore $(\varphi,\varphi) < 0$, as otherwise the linear span of $E_+(T)$ and φ would be a nonnegative subspace of dimension $\varkappa + 1$.

Corollary 2.[1]) If the expansive operator T in $\Pi_\varkappa$ has the property $\mu(T) > 1$, then $\{\zeta \mid 1 < |\zeta| < \sqrt{\mu(T)}\} \subset \rho(T)$, the invariant subspaces $\mathfrak{I}_+ \in \mathfrak{G}_+(T)$ and $\mathfrak{I}_- \in \mathfrak{G}_-(T)$ are uniquely determined, and $|\sigma(T|\mathfrak{I}_+)| \geq \sqrt{\mu(T)}$, $\mathfrak{I}_+ \dotplus \mathfrak{I}_- = \Pi_\varkappa$.

Indeed, Lemma 11.1 implies $\{\zeta \mid |\zeta| > 1\} \subset \sigma_d(T)$. Consider $\mathfrak{I}_+ \in \mathfrak{G}_+(T)$. By Theorem 11.2, it contains all the root manifolds $S_\lambda(T)$, $|\lambda| > 1$. Applying Theorem 11.2 to the expansive operator $\mu(T)^{-1/2}T$, we find a maximal nonnegative subspace $\hat{\mathfrak{I}}_+$, invariant under T and such that $|\sigma(T|\hat{\mathfrak{I}}_+)| \geq \sqrt{\mu(T)}$. Hence $\mathfrak{I}_+ \supset \hat{\mathfrak{I}}_+$, and we even have $\mathfrak{I}_+ = \hat{\mathfrak{I}}_+$, as both subspaces are $\varkappa$-dimensional. Therefore $\mathfrak{I}_+$ is the linear span of all $S_\lambda(T)$, $|\lambda| \geq \sqrt{\mu(T)}$, and the set $\{\zeta \mid 1 < |\zeta| < \sqrt{\mu(T)}\}$ cannot contain any eigenvalue of T. Applying these considerations to T^* instead of T, we see that the invariant subspaces $\mathfrak{I}_+^*$ and $\mathfrak{I}_- = (\mathfrak{I}_+^*)^{\perp}$ are also uniquely determined. The last relation of the corollary is now easy to verify.

Corollary 3. The linear span $E_-(T)$ of all root manifolds $S_\lambda(T)$, $|\lambda| < 1$, of the expansive operator T in $\Pi_\varkappa$ is a nonnegative subspace.

This follows immediately from Theorem 11.2.

---

[1])Corollary 2 also follows easily from 6.5, 2^0 and the corollary of Theorem 11.1.

We mention that all the root manifolds $S_\lambda(T)$ with $|\lambda| > 1$ belong to points $\lambda \in \sigma_d(T)$ (see Lemma 11.1). This is not necessarily true if $|\lambda| < 1$. In this case $S_\lambda(T)$ can even be infinite-dimensional. Simple examples with this property can be given using the matrix representation

$$T = \begin{bmatrix} T_{11} & 0 \\ 0 & T_{22} \end{bmatrix}$$

with respect to $\Pi_\varkappa = \Pi_+ \oplus \Pi_-$.

From the decomposition in the first part of the proof of Theorem 11.2 it follows that in case of a nonseparable space $\Pi_\varkappa$ each $\mathfrak{J}_- \in \mathfrak{S}_-(T)$ contains a (nonseparable) Hilbert space (with respect to the H-metric), which is invariant under T and T*.

Corollary 4. If the operator T is expansive, so is T*T. If, additionally, $\mu(T) > 1$, then the invariant subspaces $\mathfrak{J}_+ \in \mathfrak{S}_+(T^*T)$ and $\mathfrak{J}_- \in \mathfrak{S}_-(T^*T)$ are uniquely determined, $\mathfrak{J}_+$ is a positive, $\mathfrak{J}_-$ a negative subspace, and we have

$$\sigma(T^*T|\mathfrak{J}_+) \geqq \mu(T), \qquad \sigma(T^*T|\mathfrak{J}_-) \leqq 1 .$$

The first statement follows immediately from the corollary of Lemma 11.3'; the uniqueness of the subspaces $\mathfrak{J}_+$ and $\mathfrak{J}_-$ from Corollary 2. Moreover, if $1 < \mu < \mu(T)^2 = \mu(T^*T)$, from 6.5, 2° we have $(T^*Tx, T^*Tx) \geqq \mu(x,x) + \delta\|x\|^2$ for all $x \in \Pi_\varkappa$ and some $\delta > 0$; hence

$$(T^*Tx, T^*Tx) \geqq \delta\|x\|^2 \quad (x \in \mathfrak{J}_+) .$$

As T*T maps $\mathfrak{J}_+$ one-to-one onto itself, $\mathfrak{J}_+$ is a positive subspace. Finally, $\mathfrak{J}_-$ is the orthogonal complement of $\mathfrak{J}_+$.

Evidently, in Corollary 4 the operator T*T can be replaced by TT*.

The following theorem completes Corollary 1 in some sense.

Theorem 11.3. If the expansive operator T has no unitary eigenvalues ε ($|\varepsilon| = 1$) with neutral eigenvectors, then the invariant subspaces $\mathfrak{J}_+ \in \mathfrak{S}_+(T)$ and $\mathfrak{J}_- \in \mathfrak{S}_-(T)$ are uniquely determined, and $\mathfrak{J}_+ \dotplus \mathfrak{J}_- = \Pi_\varkappa$.

Proof. Consider an arbitrary pair of subspaces $\mathfrak{J}_+$ and $\mathfrak{J}_-$, which exist according to Theorem 11.2. Then $E_+(T) \subset \mathfrak{J}_+$ and

$$\mathfrak{J}_+ = E_+(T) \dotplus \mathfrak{J}' , \tag{11.15}$$

where $T\mathfrak{J}' \subset \mathfrak{J}'$ and $|\sigma(T|\mathfrak{J}')| = 1$. Let $\varepsilon \in \sigma(T|\mathfrak{J}')$ and $\varphi_o \in \mathfrak{J}'$, $T\varphi_o = \varepsilon\varphi_o$ ($\varphi_o \neq \theta$). Then $\varphi_o \in \mathfrak{J}_+$, whence $(\varphi_o, \varphi_o) \geqq 0$; here the sign = is excluded by the condition of the theorem. Denote by $L_\varepsilon(T)$ again the eigenspace of T corresponding to the eigenvalue ε. As it cannot contain neutral vectors it must be definite, and $\varphi_0 \in L_\varepsilon(T)$ implies that $L_\varepsilon(T)$ is positive. Hence $L_\varepsilon(T) = S_\varepsilon(T)$ by Lemma 11.6.

Denote by ε_j, $j = 1,\dots,p$, the unitary eigenvalues of T for which the corresponding eigenspaces $L_{\varepsilon_j}(T) = S_{\varepsilon_j}(T)$, $j = 1,\dots,p$, are positive. According to Lemma 11.6 they are mutually orthogonal and orthogonal to $E_+(T)$. The direct sum

$$\mathfrak{J}_+^0 = E_+(T) \dotplus L_{\varepsilon_1}(T) \dotplus L_{\varepsilon_2}(T) \dotplus \dots \dotplus L_{\varepsilon_p}(T) \tag{11.16}$$

$$(= E_+(T) \dotplus S_{\varepsilon_1}(T) \dotplus S_{\varepsilon_2}(T) \dotplus \dots \dotplus S_{\varepsilon_p}(T))$$

is therefore a nonnegative subspace, invariant under T. From (11.15) and the inclusion $\sigma(T|\mathfrak{J}') \subset \{\varepsilon_1,\dots,\varepsilon_p\}$ it follows that $\mathfrak{J}_+^0 \supset \mathfrak{J}_+$, and hence $\mathfrak{J}_+ = \mathfrak{J}_+^0$.

The unitary eigenvalues of T^* have the same properties as those of T (see Lemma 3.9), and $\bar\varepsilon_1,\dots,\bar\varepsilon_p$ are the unitary eigenvalues of T^* with positive eigenspaces $L_{\bar\varepsilon_j}(T^*) = S_{\bar\varepsilon_j}(T^*) = L_{\varepsilon_j}(T)$, $j = 1,\dots,p$. Therefore $\mathfrak{S}_+(T^*)$ contains only one subspace, which is

$$\mathfrak{J}_+^* = E_+(T^*) \dotplus L_{\varepsilon_1}(T) \dotplus \dots \dotplus L_{\varepsilon_p}(T) ;$$

hence the unique element of $\mathfrak{S}_-(T)$ is $\mathfrak{J}_- = (\mathfrak{J}_+^*)^\perp$.

It is easy to see that $\mathfrak{J}_- \cap \mathfrak{J}_+^* = \{\theta\} = \mathfrak{J}_- \cap \mathfrak{J}_+$. The first equality implies $\mathfrak{J}_- \oplus \mathfrak{J}_+^* = \Pi_\varkappa$, and from $\dim \mathfrak{J}_+ = \dim \mathfrak{J}_+^* = \varkappa$ we have, finally, $\mathfrak{J}_- \dotplus \mathfrak{J}_+ = \Pi_\varkappa$.

The theorem is proved.

In the course of the proof of Theorem 3.3 we obtained a description of the subspace $\mathfrak{J}_+ = \mathfrak{J}_+^0$, (see (11.16)).

The conditions of Theorem 3.3 are satisfied, e.g., if T is a B-plus operator (see Definition 2.9).

It is possible that for the strict plus-operator T either of the sets $\mathfrak{S}_\pm(T)$ consists only of one subspace, but T has other maximal nonnegative and maximal nonpositive invariant subspaces $\mathfrak{J}_+^!$ and $\mathfrak{J}_-^!$ respectively. Then

$$\sigma(T|\mathfrak{J}_+^!) \cap \{\zeta \mid |\zeta| < 1\} \neq \emptyset \quad \text{and} \quad \sigma(T|\mathfrak{J}_-^!) \cap \{\zeta \mid |\zeta| > 1\} \neq \emptyset .$$

This happens e.g. for a unitary operator U with a nontrivial so-called hyperbolic invariant subspace (see § 12.2). Other examples are given by operators which are "close" to unitary operators with this property. In this connection the following result may be of some interest. Here $E_+(T)$ denotes again the linear span of all root manifolds $S_\lambda(T)$, $|\lambda| > 1$, of the expansive operator T.

<u>Theorem 11.4.</u> Let T be an expansive operator in $\Pi_\varkappa$ with the property

$$(Tx,Tx) > (x,x) \quad \text{for} \quad (x,x) \geqq 0,\ x \neq \theta . \tag{11.17}$$

Then also

$$(T^*x,T^*x) > (x,x) \quad \text{for} \quad (x,x) \geqq 0,\ x \neq \theta , \tag{11.18}$$

and $\mathfrak{I}_+ = E_+(T)$ ($\mathfrak{I}_- = E_+(T^*)^\perp$) is the unique $\varkappa$-dimensional nonnegative (maximal nonpositive) subspace of $\Pi_\varkappa$ which is invariant under T. Moreover, $\mathfrak{I}_+$ ($\mathfrak{I}_-$) is even positive (negative respectively), and $\Pi_\varkappa = \mathfrak{I}_+ \dot{+} \mathfrak{I}_-$.

<u>Proof.</u> From Lemma 11.3 we find that $(T^*x,T^*x) \geqq (x,x)$ $(x \in \Pi_\varkappa)$. Assume $(T^*x_o,T^*x_o) = (x_o,x_o)$ for some $x_o \in \Pi_\varkappa$, $(x_o,x_o) \geqq 0$. Then (see Lemma 11.5) $TT^*x_o = x_o$ and $T^*T(T^*x_o) = T^*x_o$. It follows that

$$(TT^*x_o,TT^*x_o) = (T^*x_o,T^*x_o) .$$

As $(T^*x_o,T^*x_o) \geqq 0$, condition (11.17) implies $T^*x_o = \theta$, and we get, finally, $x_o = \theta$; that is (11.18) holds.

Consider an arbitrary $\varkappa$-dimensional nonnegative subspace $\mathfrak{I}_+$ invariant under T. Then (11.17) implies $|\sigma(T|\hat{\mathfrak{I}}_+)| > 1$; hence $\hat{\mathfrak{I}}_+ \in E_+(T)$. But the subspace $E_+(T)$ is at most $\varkappa$-dimensional; therefore $\hat{\mathfrak{I}}_+ = E_+(T)$. Moreover, from (11.17) and $T\hat{\mathfrak{I}}_+ = \hat{\mathfrak{I}}_+$ it follows that $\hat{\mathfrak{I}}_+ = E_+(T)$ is positive.

If $\hat{\mathfrak{I}}_-$ is a maximal nonpositive subspace, invariant under T, then $\hat{\mathfrak{I}}_-^\perp$ is a $\varkappa$-dimensional nonnegative subspace of $\Pi_\varkappa$, invariant under T^*. As T^* satisfies the same conditions as T (see (11.18)), we find that $\hat{\mathfrak{I}}_-^\perp = E_+(T^*)$ and that this is a positive subspace. Therefore $E_+(T^*)^\perp$ is the unique maximal nonpositive invariant subspace of T, and, being the orthogonal complement of a positive subspace, it is even negative. The decomposition $\Pi_\varkappa = E_+(T) \dot{+} E_+(T^*)$ now follows e.g. from Corollary 1 of Theorem 11.2.

In the same way as the corollary of Theorem 11.1 gives more specific information in the case of a selfadjoint uniformly expansive operator, Theorems 11.2 - 4 have special conclusions for selfadjoint expansive operators. Their formulation will be left to the reader.

11.5. Definition 11.1. A continuous linear operator T in $\Pi_\varkappa$ $(D(T) = \Pi_\varkappa)$ is called contractive if $(Tx,Tx) \leqq (x,x)$ $(x \in \Pi_\varkappa)$.

The adjoint T^* of a contractive operator T in $\Pi_\varkappa$ is not necessarily contractive (whereas the corresponding statement for an expansive operator holds - see Lemma 11.3). This can be seen from the example

$$T = \begin{bmatrix} 0 & 0 \\ 0 & T_{22} \end{bmatrix} \quad \text{in } \Pi_\varkappa = \Pi_+ \oplus \Pi_- ,$$

where T_{22} $(D(T_{22}) = \Pi_-)$ is an isometric, nonunitary operator in Π_-. If T and T^* are contractive, T is called doubly contractive.

Lemma 11.7. If T is contractive and α has the properties $|\alpha| < 1$ and $\overline{\alpha}^{-1} \notin \sigma(T)$, then the operator $T_\alpha = (T - \alpha I)(I - \overline{\alpha}T)^{-1}$ is contractive. If, moreover, also $\alpha \notin \sigma(T)$, then T_α is fully invertible; that is, $0 \in \rho(T_\alpha)$.

Proof. The relation $(T_\alpha x, T_\alpha x) \leqq (x,x)$, $x \in \Pi_\varkappa$, which we have to prove, is equivalent to

$$((T - \alpha I)y,\ (T - \alpha I)y) \leqq ((I - \overline{\alpha}T)y,\ (I - \overline{\alpha}T)y), \qquad y = (I - \overline{\alpha}T)^{-1}x .$$

This can be written as

$$(1 - |\alpha|^2)(Ty,Ty) \leqq (1 - |\alpha|^2)(y,y) ,$$

and the statement follows.

Lemma 11.8 (cf. Lemma 11.1). Let T be a contractive operator in $\Pi_\varkappa$. Then the set $D = \{\zeta \mid |\zeta| < 1\}$, with the possible exception of an at most countable number of isolated points in D, consists of points of regular type for T. Denoting this set of points of regular type by D_T, we have

$$n(\zeta) \equiv \dim((T - \zeta I)\Pi_\varkappa)^\perp = n_0 \text{ on } D_T ;$$

at the exceptional points $\zeta \in D \setminus D_T$ the inequality $n(\zeta) > n_0$ holds.

Proof. Let $T = \|T_{ij}\|^2_{i,j=1}$ be the matrix representation of T with respect to the canonical decomposition (1.1). Then we have

$$T_{22}^*T_{22} \geqq I_- + T_{12}^*T_{12} ;$$

therefore the operator T_{22} is of the form $T_{22} = S_2 + R_2$ where $S_2^*S_2 \geqq I_-$ and R_2 is a finite-dimensional operator (cf. the proof of Lemma 11.1). Consider the decompositon

$$T = \begin{bmatrix} T_{11} - I_+ & T_{12} \\ T_{21} & R_2 \end{bmatrix} + \begin{bmatrix} I_+ & 0 \\ 0 & S_2 \end{bmatrix} = T_1 + T_2 .$$

It is easy to see that D belongs to the set of points of regular type of $T_2 = \begin{bmatrix} I_+ & 0 \\ 0 & S_2 \end{bmatrix}$. As the operator T_1 is finite-dimensional, the statement follows from [17a].

Lemma 11.9. Let T be a contractive operator in $\Pi_\varkappa$. Then the following statements are equivalent:

(1) There exists at least one point $\alpha \in \rho(T)$, $|\alpha| < 1$.

(2) The operator T is doubly contractive.

(3) For at least one canonical decomposition $\Pi_\varkappa = \Pi_+ \oplus \Pi_-$ the operator $P_-T|\Pi_-$ is fully invertible.

(4) For all canonical decompositions $\Pi_\varkappa = \Pi_+ \oplus \Pi_-$ the operator $P_-T|\Pi_-$ is fully invertible.

(5) T maps each maximal nonpositive subspace onto a maximal nonpositive subspace.

Proof. Suppose (1) holds. Then, with the notation of Lemma 11.8, we have $n(\zeta) = 0$ if $\zeta \in D_T$; hence there exists an $\alpha \in \rho(T)$, $|\alpha| < \min(1, \|T\|^{-1})$, and by Lemma 11.7 the operator T_α is contractive and fully invertible. Obviously T_α^{-1} is expansive, and by Lemma 11.3 the same is true for $(T_\alpha^{-1})^* = (T_\alpha^*)^{-1}$, which means that T_α^* is contractive. Then, again by Lemma 11.7, the same is true for $T^* = (T_\alpha^* + \bar{\alpha} I)(I + \alpha T_\alpha^*)^{-1}$; that is, (2) holds. These considerations can be reversed, and the equivalence of (1) and (2) follows.

If T is doubly contractive, for all canonical decompositions $\Pi_\varkappa = \Pi_+ \oplus \Pi_-$ we have for $T_{22} = P_-T|\Pi_-$, that $T_{22}^* T_{22} \geqq I_-$ and $T_{22} T_{22}^* \geqq I_-$. The first inequality implies that $\zeta = 0$ is a point of regular type of T_{22}, and from the second inequality it follows that the range of T_{22} is the whole space Π_-. Hence (2) implies (3) and (4).

Assume now that (3) holds. Then also the operator $P_+ + P_-T$ is fully invertible[1]). The hypothesis that T is contractive can be written as

[1]) This follows immediately from the matrix representation

$$P_+ + P_-T = \begin{bmatrix} I_+ & 0 \\ T_{21} & T_{22} \end{bmatrix}$$

and the condition that T_{22} is fully invertible.

$$\|P_+Tx\|^2 - \|P_-Tx\|^2 \leq \|P_+x\|^2 - \|P_-x\|^2 \quad (x \in \Pi_\varkappa) ,$$

or

$$\|(P_+T + P_-)x\|^2 = \|P_+Tx\|^2 + \|P_-x\|^2 \leq \|P_-Tx\|^2 + \|P_+x\|^2 =$$
$$= \|(P_-T + P_+)x\|^2 \quad (x \in \Pi_\varkappa) ;$$

hence T is contractive in $\Pi_\varkappa$ if and only if the operator

$$\Psi(T) = (P_+T + P_-)(P_-T + P_+)^{-1} \tag{11.19}$$

is contractive with respect to the H-norm $\|x\|^2 = (P_+x,P_+x) - (P_-x,P_-x)$ on $\Pi_\varkappa$. Then the same is true for $\Psi(T)^*$; that is, we have

$$\|(T^*P_- + P_+)^{-1} (T^*P_+ + P_-)\| \leq 1 .$$

Now it can be checked directly that the equality

$$-(T^*P_- + P_+)^{-1}(T^*P_+ + P_-) = (P_- - P_+T^*)(P_+ - P_-T^*)^{-1}$$

holds; hence the right-hand side is also a contractive operator with respect to this H-norm. By **the considerations of the first part of this** paragraph, $-T^*$ and therefore also T^* are contractive in $\Pi_\varkappa$; that is, T is doubly contractive.

The equivalence of (3) and (4) is now obvious.

Suppose again that (2) holds, and let L' be a maximal nonpositive subspace with angle operator $K'|\Pi_- \to \Pi_+$, $\|K'\| \leq 1$ (see 10.1). As the range of the operator $T_{21}K' + T_{22}$ is closed (note again that $T_{21}K'$ is a finite-dimensional operator), the statement (5) follows if we only show that this range is the whole space Π_- (cf. Lemma 10.2). Assume, to the contrary, that there exists a nonzero vector $y_- \in \Pi_-$ such that

$$(K'^*T_{21}^* + T_{22}^*)y_- = \theta . \tag{11.20}$$

As T is doubly contractive, we have $(T^*y_-,T^*y_-) \leq (y_-,y_-)$ or

$$\|T_{21}^*y_-\|^2 \leq \|T_{22}^*y_-\|^2 - \|y_-\|^2 < \|T_{22}^*y_-\|^2 ,$$

and from (11.20) we get

$$\|T_{22}^*y_-\| = \|K'^*T_{21}^*y_-\| \leq \|T_{21}^*y_-\| < \|T_{22}^*y_-\| ,$$

a contradiction. Hence (2) implies (5). On the other hand, it is easy to see that (5) implies (3).

To complete the proof of the lemma we show that (2) and (4) imply (1). Put $L_0^* = \{\varphi | T^*\varphi = \theta\}$. If $L_0^* = \{\theta\}$, it follows from Lemma 11.8 that T is fully invertible. If $L_0^* \neq \{\theta\}$, then it can be shown as in the proof of Corollary 1 to Lemma 11.5 that L_0^* is a positive subspace. Then $(L_0^*)^{\perp} = \Pi'_{\varkappa'}$ with index $\varkappa' = \varkappa - \dim L_0^*$, and

$$T\Pi'_{\varkappa'} \subset \Pi'_{\varkappa'} . \tag{11.21}$$

The operator $T' = T|\Pi'_{\varkappa'}$ in $\Pi'_{\varkappa'}$ has again the properties (2) and (4). If in (11.21) the sign $\subsetneqq$ holds, the same reasoning can be repeated with T' instead of T, etc., giving after at most $\varkappa$ steps a subspace $\hat{\Pi}_{\hat{\varkappa}}$ of $\Pi_\varkappa$ such that $(\hat{\Pi}_{\hat{\varkappa}})^{\perp}$ is positive and $T\hat{\Pi}_{\hat{\varkappa}} = \hat{\Pi}_{\hat{\varkappa}}$ because of condition (4). Then the matrix representation of T with respect to the decomposition $\Pi_\varkappa = (\hat{\Pi}_{\hat{\varkappa}})^{\perp} \oplus \hat{\Pi}_{\hat{\varkappa}}$ is

$$T = \begin{bmatrix} T_1 & 0 \\ \cdot & T_2 \end{bmatrix} .$$

Here T_2 is fully invertible and T_1 is a finite-dimensional operator. Evidently, in an arbitrary neighbourhood of zero there are points of $\rho(T_1) \cap \rho(T_2) = \rho(T)$.

Lemma 11.9 is proved.

As an immediate consequence of Lemma 11.9, Lemma 11.7, Theorem 11.2 and Theorem 11.4 we have now the following result.

<u>Theorem 11.5.</u> Let T be a contractive operator in $\Pi_\varkappa$ with the properties (1) - (5) of Lemma 11.9. Then the following statements hold:

(i) All the points $\lambda \in \sigma(T)$, $|\lambda| < 1$, belong to $\sigma_d(T)$.

(ii) There exists a maximal nonnegative (nonpositive) subspace $\mathfrak{I}_+$ ($\mathfrak{I}_-$) which is invariant under T and has the property $|\sigma(T|\mathfrak{I}_+)| \leqq 1$ ($|\sigma(T|\mathfrak{I}_-)| \geqq 1$, respectively).

(iii) Each subspace $\mathfrak{I}_+$ ($\mathfrak{I}_-$) with the properties given in (ii) contains all the root subspaces $S_\lambda(T)$ with $|\lambda| < 1$ ($|\lambda| > 1$).

(iv) If T is properly contractive (that is, $(Tx,Tx) < (x,x)$ for all $x \in \Pi_\varkappa \setminus \{\theta\}$), then there is exactly one maximal nonnegative (maximal nonpositive) subspace $\mathfrak{I}_+$ ($\mathfrak{I}_-$) of $\Pi_\varkappa$ which is invariant under T; this subspace is even positive (negative, respectively), and $\Pi_\varkappa = \mathfrak{I}_+ \dotplus \mathfrak{I}_-$.

The formulation of other immediate consequences from the results of 11.3 is left to the reader.

In analogy to uniformly expansive operators the continuous linear operator T is called <u>uniformly contractive</u> if for some $\varepsilon > 0$ we have

$(Tx,Tx) \leqq (x,x) - \varepsilon\|x\|^2 \quad (x \in \Pi_\varkappa)$. It is easy to see that the contractive operator T is uniformly contractive if and only if the operator $\Psi(T)$ in (11.19) is a uniform contraction with respect to the H-norm: $\|\Psi(T)\| < 1$.

If the operator T is uniformly contractive, $|\alpha| < 1$ and $\bar{\alpha}^{-1} \in \rho(T)$, then the operator $T_\alpha = (T - \alpha I)(I - \bar{\alpha}T)^{-1}$ is also uniformly contractive; this can be checked easily. Moreover, if also $\alpha \in \rho(T)$, then T_α is fully invertible and T_α^{-1} is uniformly expansive. Then, according to Lemma 11.2, for some $\delta > 0$ the set $\rho(T_\alpha^{-1})$ contains the ring $\Gamma_\delta = \{\zeta \mid (1 + \delta)^{-1} < |\zeta| < 1 + \delta\}$; hence $\rho(T_\alpha)$ contains the same ring Γ_δ. Then, by virtue of the relation $T = (T_\alpha + \alpha I)(I + \bar{\alpha}T_\alpha)^{-1}$, the set $\rho(T)$ contains some ring $\Gamma_{\delta'}$, $\delta' > 0$. Thus, one direction of the following statement (analogue of Lemma 11.2) is proved.

A uniformly contractive operator T in $\Pi_\varkappa$ is doubly contractive if and only if the unit circle belongs to $\rho(T)$.

The sufficiency can be proved in the same way as the corresponding part in Lemma 11.3.

11.6. Definition 11.2. A linear operator B in $\Pi_\varkappa$ with dense domain D(B) is called dissipative if $\operatorname{Im}(Bx,x) \geqq 0$ for all $x \in D(B)$, and maximal dissipative if it does not have a proper extension in $\Pi_\varkappa$ which is a dissipative operator.

We recall that a linear operator A, $\overline{D(A)} = H$, in a Hilbert space H with scalar product $[x,y]$ is called dissipative (in our terminology H-dissipative) if $\operatorname{Im}[Ax,x] \geqq 0$ for $x \in D(A)$, and maximal dissipative if it does not have proper dissipative extensions. According to a well-known theorem of R.S. Phillips [34a], for a dissipative operator A the lower half plane consists only of points of regular type, and A is maximal dissipative if and only if at least one point (and then all the points) of the lower half plane belongs to $\rho(A)$.

Returning to dissipative operators in $\Pi_\varkappa$, we again choose a canonical decomposition (1.1) of $\Pi_\varkappa$, such that $\Pi_+ \subset D(B)$, and the corresponding operator J (see 3.1). Then B is dissipative (maximal dissipative) if and only if JB is H-dissipative (maximal H-dissipative); that is, $\operatorname{Im}[JBx,x] \geqq 0$ for $x \in D(B) = D(JB)$; in this statement the operator JB can be replaced by the operator BJ (both operators are unitarily equivalent). From this correspondence between dissipative and H-dissipative operators and well-known results about H-dissipative operators we have the following results.

1°. Each dissipative operator in $\Pi_\varkappa$ has a closure which is again a dissipative operator.

2°. Each dissipative operator in $\Pi_\varkappa$ has an extension in $\Pi_\varkappa$ which is a maximal dissipative operator.

3°. If B is maximal dissipative, so is $-B^*$.

Writing $-B = +BJ - 2BP_+$, where $J = P_+ - P_-$ and P_+ (P_-) is the orthoprojector on Π_+ (Π_-), we see that for dissipative B the operator $-B$, is the sum of an H-dissipative and a finite-dimensional operator. From the perturbation type argument used several times above we immediately get

4°. If B is a maximal dissipative operator in $\Pi_\varkappa$, the spectrum of B in the open upper half plane belongs to $\sigma_d(B)$.

The above mentioned criterion of R.S. Phillips for the maximality of a dissipative operator has the following analogue in $\Pi_\varkappa$:

5°. A dissipative operator B in $\Pi_\varkappa$ is maximal dissipative if and only if in the upper half plane there is at least one point of $\rho(B)$.

The proof of the following lemma is similar to the proofs of the corresponding propositions in 8.1, and can be left to the reader.

Lemma 11.10. Let B be a densely defined linear operator in $\Pi_\varkappa$, and let $\zeta \in \rho(B)$, $\mathrm{Im}\,\zeta > 0$. Then B is maximal dissipative if and only if the Cayley transform $T = (B - \bar{\zeta}I)(B - \zeta I)^{-1}$ is an expansive operator.

In order to simplify the formulation of the following theorem we shall say that a subspace L is invariant under the linear operator B ($\overline{D(B)} = \Pi_\varkappa$) if $\overline{L \cap D(B)} = L$ and $B(L \cap D(B)) \subset L$. In this case the restriction of B to $L \cap D(B)$ will be denoted by $B|L$.

The theorems, lemmas and corollaries proved in 11.3 and 11.4 and the properties of the Cayley transformation now imply

Theorem 11.6. Let B be a maximal dissipative operator in $\Pi_\varkappa$. Then the following statements hold:

(i) There exists a maximal nonnegative (nonpositive) subspace $\mathfrak{J}_+$ ($\mathfrak{J}_-$) which is invariant under B and such that $\mathrm{Im}\,\sigma(B|\mathfrak{J}_+) \geqq 0$ ($\mathrm{Im}\,\sigma(B|\mathfrak{J}_-) \leqq 0$).

(ii) Each subspace $\mathfrak{J}_+$ ($\mathfrak{J}_-$) with the properties in (i) contains every root manifold $S_\lambda(B)$ with $\mathrm{Im}\,\lambda > 0$ ($\mathrm{Im}\,\lambda < 0$).

(iii) If B is, moreover, properly dissipative (that is, $\mathrm{Im}(Bx,x) > 0$ for $x \in D(B) \setminus \{\theta\}$), then there exists a unique maximal nonnegative (nonpositive) subspace $\mathfrak{J}_+$ ($\mathfrak{J}_-$) invariant under B; this subspace is even positive (negative), and $\mathfrak{J}_+ \dotplus \mathfrak{J}_- = \Pi_\varkappa$.

From (ii) it is evident that the maximal dissipative operator B has in the open upper half plane at most $\varkappa$ eigenvalues, each counted according to its algebraic multiplicity.

The formulation of other consequences for maximal dissipative operators of the results in 11.3 and 11.4 will be left to the reader.

Historical Notes to § 11.

1. The existence of a $\varkappa$-dimensional nonnegative invariant subspace for a given expansive operator T in $\Pi_\varkappa$ was first proved by M.G. Krein in [21a] as an application of Brower's fixed point principle.[1] At that time this fact was apparently new even in the algebraic case ($\dim \Pi_\varkappa < \infty$; see [21b]). Later M.L. Brodskiĭ [10] generalized this result to operators T in a reflexive Banach space with indefinite metric and investigated the spectrum of T (some of his results were generalized to arbitrary (nonreflexive) Banach spaces by H. Langer [28c]). For an expansive operator T in an arbitrary space Π with regular indefinite metric the existence of a maximal nonnegative invariant subspace can be proved by applying a fixed-point principle, if only the operator T_{12} is compact (see e.g. [21f]); a compactness condition of this kind first appeared in [28a, b]).

Let us sketch such a proof in case of a space $\Pi_\varkappa$ (then T_{12} is a finite-dimensional operator). If L is a $\varkappa$-dimensional nonnegative subspace and K is its angle operator, then TL is again a $\varkappa$-dimensional nonnegative subspace and $T_{11} + T_{12}K$ is invertible. Further, Lemma 10.3 and (10.6) imply that the $\varkappa$-dimensional nonnegative subspace L_0 is invariant under T if and only if its angle operator K_0 is a fixed point of the mapping

$$K \to \Phi(K) \equiv (T_{21} + T_{22}K)(T_{11} + T_{12}K)^{-1} .$$

The set $\mathfrak{K}$ of all operators K from Π_+ to Π_-, $\|K\| \leq 1$, is convex and compact in the weak operator topology. Moreover, it can be shown that the mapping Φ is continuous in this topology. Hence in case of a

[1]) Compare also the Historical Notes to Theorems 12.1 and 12.1' in 12.1.

separable space $\Pi_\varkappa$ Schauder's fixed-point principle implies the existence of a fixed point K_0 of Φ, and the corresponding $\varkappa$-dimensional nonnegative subspace L_0 is invariant under T_0.

2. Theorem 11.1 was first proved in the Supplement of the 1956 edition of this article. As we have already mentioned, in its general form in spaces Π with a regular indefinite metric it plays an essential role in the stability theory of differential equations in Hilbert space [11]. Here we use it to prove the existence of a $\varkappa$-dimensional nonnegative invariant subspace of an expansive operator T (see Theorem 11.2) through approximation of T by a sequence of uniformly expansive operators $T(\delta)$. Thus we are able to avoid any fixed-point principle.

Theorem 11.2 is the central result of § 11 and is published in this form here for the first time. However, its first statement as well as the inclusion $S_\lambda \subset \mathfrak{J}_+$ for at least one $\varkappa$-dimensional nonnegative invariant subspace $\mathfrak{J}_+$ of T in their equivalent form for a maximal dissipative operator (see Theorem 11.6) were proved by M.G. Kreĭn and H. Langer [24d] and independently by T.Ja. Azizov [3b]. To Azizov the authors owe also the main statements of Lemmas 11.5 and 11.6, which play an essential role in the proof of Theorems 11.2 - 4. In its variant for maximal dissipative operators Theorem 11.3 was proved by Azizov [3b] (in a slightly weaker form). The simple Theorem 11.4 is published here for the first time. This is also true for the main results of 11.5 on contractive operators. However, after the crucial Lemma 11.9 has been proved, they are immediate consequences of the foregoing results on expansive operators. As regards Lemma 11.9, it is close to earlier results of Ju.P. Ginzburg [16a] and M.G. Kreĭn and Ju. L. Šmul'jan [26a] for arbitrary spaces Π with regular indefinite metric. The fractional linear transformation (11.19), which gives a one-to-one correspondence between the set of contractive operators in $\Pi_\varkappa$ and some set of H-contractive operators, was in the finite-dimensional case first considered by V.P. Potapov [36] and in the general case of a space Π by Ju.P. Ginzburg [16a] (see also I.S. Iohvidov [18g], where a reference to an earlier paper by Bott can be found); Ginzburg proved the equivalence of (2) and (3) in this general situation.

§ 12. Invariant semidefinite subspaces of unitary and selfadjoint operators in $\Pi_\varkappa$

12.1. Below we formulate some propositions in parallel for unitary and selfadjoint operators. Sometimes the proofs are given only for unitary operators; they are completely analogous to the corresponding proofs by L.S. Pontrjagin for selfadjoint operators. Each of these propositions can be obtained from the "dual" proposition, formulated in parallel, by the Cayley transformation (see § 8, Theorems 8.2 and 8.2').

Theorem 12.1. Let U be a unitary operator in $\Pi_\varkappa$. Then there are $\varkappa$-dimensional nonnegative subspaces $\mathfrak{J}_+$ and $\mathfrak{J}_+'$, invariant under U and such that $|\sigma(U|\mathfrak{J}_+)| \geqq 1$ and $|\sigma(U|\mathfrak{J}_+')| \leqq 1$. Moreover, $\mathfrak{J}_+$ ($\mathfrak{J}_+'$) contains all the root manifolds $S_\lambda(U)$ belonging to eigenvalues λ with $|\lambda| > 1$ ($|\lambda| < 1$).

Proof. As the unitary operator U satisfies the conditions of both Theorems 11.2 and 11.5, the assertions of Theorem 12.1 follow immediately.

Corollary. A unitary operator U in $\Pi_\varkappa$ has at least one nonnegative eigenvector.

Note. The orthogonal complements $\mathfrak{J}_- = \mathfrak{J}_+^\perp$ and $\mathfrak{J}_-' = (\mathfrak{J}_+')^\perp$ are maximal nonpositive subspaces. It is easy to see that they are invariant under U. Moreover, as $S_\lambda(U)$, $|\lambda| \neq 1$, is neutral (see Theorem 7.1), they have the properties $|\sigma(U|\mathfrak{J}_-)| \geqq 1$ and $|\sigma(U|\mathfrak{J}_-')| \leqq 1$, and $\mathfrak{J}_-$ ($\mathfrak{J}_-'$) contains all the root manifolds $S_\lambda(U)$ belonging to eigenvalues λ with $|\lambda| > 1$ ($|\lambda| < 1$).

Actually, Theorems 11.2 and 11.5 allow us to formulate even a somewhat stronger statement than Theorem 12.1, but this will be left to the reader.

If we are given a selfadjoint operator A in $\Pi_\varkappa$ for which the nonreal numbers ζ and $\bar{\zeta}$ are not eigenvalues, then according to 8.1 the operator U, defined by

$$U = (A - \bar{\zeta}I)(A - \zeta I)^{-1} ,$$

is a unitary operator in $\Pi_\varkappa$. From this we obtain

Theorem 12.1'. Let A be a selfadjoint operator in $\Pi_\varkappa$. Then D(A) contains $\varkappa$-dimensional nonnegative subspaces $\mathfrak{J}_+$ and $\mathfrak{J}_+'$, invariant under A and such that $\operatorname{Im} \sigma(A|\mathfrak{J}_+) \geqq 0$ and $\operatorname{Im} \sigma(A|\mathfrak{J}_+') \leqq 0$. Moreover $\mathfrak{J}_+$ ($\mathfrak{J}_+'$) contains all the root manifolds $S_\lambda(A)$ belonging to eigenvalues λ with $\operatorname{Im} \lambda > 0$ ($\operatorname{Im} \lambda < 0$).

The proof follows immediately from Theorems 12.1, 8.2 and 8.2'.

Corollary. A selfadjoint operator A in $\Pi_\varkappa$ has at least one nonnegative eigenvector.

Note. With $\mathfrak{J}_+$ and $\mathfrak{J}_+'$ from Theorem 12.1', the subspaces $\mathfrak{J}_- = \mathfrak{J}_+^\perp$ and $\mathfrak{J}_-' = (\mathfrak{J}_+')^\perp$ are maximal nonpositive, invariant[1]) under the selfadjoint operator A and have the property that $\operatorname{Im} \sigma(A|\mathfrak{J}_-) \geqq 0$ and $\operatorname{Im} \sigma(A|\mathfrak{J}_-') \leqq 0$. Moreover, $\mathfrak{J}_-$ ($\mathfrak{J}_-'$) contains all the root manifolds $S_\lambda(A)$ belonging to eigenvalues λ with $\operatorname{Im} \lambda > 0$ ($\operatorname{Im} \lambda < 0$).

Historical Notes to Theorems 12.1 and 12.1'.

Theorem 12.1' was first proved by L.S. Pontrjagin [35] in a completely different way from ours. Before that the special case $\varkappa = 1$ had been treated by S.L. Sobolev in connection with a problem of mechanics, but the corresponding publication [38] appeared only in 1960. The fundamental paper [35] was the starting-point of the investigations of linear operators in $\Pi_\varkappa$.

Independently of Pontrjagin's work, M.G. Kreĭn [25] proved Theorem 12.1 for a real space Π_1 in connection with his theory of operators in Banach spaces with cone. Then I.S. Iohvidov in 1946 (see [18b]) deduced it in its full generality from Pontrjagin's Theorem 12.1', applying for the first time the Cayley transformation to operators in spaces with indefinite metric. Three years later Kreĭn [21a] proved it using a fixed-point principle (see the Historical Notes to § 11).

In this paper Theorem 12.1 appears as a consequence of Theorem 11.2 by means of the Cayley transformation. This method was used by Kreĭn [21f] in case of a general space Π with regular indefinite metric. He showed the existence of a maximal nonnegative invariant subspace of a selfadjoint operator A in Π, if only $A_{12}(A_{22} - \lambda I)^{-1}$ is compact for some $\lambda \in \rho(A_{22})$. Earlier H. Langer [28a,b] proved this statement under

[1])In the sense explained before Theorem 11.6.

the stronger[1]) condition of the compactness of A_{12}.

12.2. Definition 12.1. The direct sum L of all the root subspaces S_λ of a unitary operator U (or a selfadjoint operator A) in $\Pi_\varkappa$ corresponding to eigenvalues λ, $|\lambda| \neq 1$ ($\lambda \neq \bar{\lambda}$, respectively) is called the hyperbolic invariant subspace of U (A).

The subspace L is finite-dimensional (see the corollary of Theorem 7.3).

Lemma 12.1 (12.1').[2]) The scalar product does not degenerate on the hyperbolic invariant subspace L of a unitary operator U (of a selfadjoint operator A) in $\Pi_\varkappa$.

Proof. Evidently (see Theorem 12.1) $L = \Lambda \dotplus \Lambda'$, where $\Lambda = L \cap \mathfrak{J}_+$ and $\Lambda' = L \cap \mathfrak{J}'_+$. The subspace Λ does not contain any nonzero vector which is orthogonal to Λ'. In fact if $x \in S_\lambda$ $(x \neq \theta)$ and $x \perp \Lambda'$, then $x \perp \mathfrak{J}'_+$, because $\mathfrak{J}'_+$ is the direct sum of Λ' and some root manifolds $S_\lambda(U)$ corresponding to eigenvalues λ, $|\lambda| = 1$. Thus $x \in \mathfrak{J}'_+$, $x \perp \mathfrak{J}'_+$ and $(x,x) = 0$. Now in this case it follows that the linear span of the subspace $\mathfrak{J}'_+$ and the vector x would be a nonnegative $(\varkappa+1)$-dimensional subspace, which is impossible (Lemma 1.2). By analogous methods it can be shown that Λ' contains no nonzero vector which is orthogonal to Λ. So Λ and Λ' are skewly linked.

As each of the subspaces Λ and Λ' is neutral, it follows from what has been proved that the scalar product does not degenerate on $L = \Lambda \dotplus \Lambda'$.

12.3. We now use as a starting point the decomposition of L into a direct sum $L = \Lambda \dotplus \Lambda'$, which was established above. It is clear that the dimension of L is not greater than $2\varkappa$.

Theorem 12.2 [18a]. For any choice of a basis in Λ we can choose a basis in Λ' in such a way that the unitary operator U, when written in

[1])Lately, however, T.Ja. Azizov and T.V. Kondras have shown that under M.G. Kreĭn's condition (compactness of $A_{12}(A_{22} - \lambda I)^{-1}$) there exists a decomposition $\Pi = \Pi_+ \oplus \Pi_-$ such that the corresponding component A_{12} is compact.

[2])This lemma as well as Theorem 12.2 and its corollaries can be proved without using the fundamental Theorems 12.1 or 12.1' (see e.g. [9]).

matrix form as an operator on the hyperbolic invariant subspace L, has the form of a block matrix

$$U|L = \begin{bmatrix} V & 0 \\ 0 & (V^*)^{-1} \end{bmatrix} ,$$

where V is a matrix acting in Λ and $(V^*)^{-1}$ is a matrix acting in Λ'.

Theorem 12.2' [35]. For any choice of a basis in Λ we can choose a basis in Λ' in such a way that the selfadjoint operator A when written in matrix form as an operator on the hyperbolic invariant subspace L takes the form of a block matrix

$$A|L = \begin{bmatrix} B & 0 \\ 0 & B^* \end{bmatrix} ,$$

where B is a matrix acting in Λ and B^* is a matrix acting in Λ'.

Proof of Theorem 12.2. Let $e_1,\dots,e_r$ be any basis of Λ and $f_1,\dots,f_s$ any basis of Λ'. It was proved above that there is no vector in Λ orthogonal to Λ' and none in Λ' orthogonal to Λ (see the proof of Lemma 12.1). From this, on the basis of elementary theorems of linear algebra it follows that the matrix

$$\| (e_j, e_k) \|_{j=1,k=1}^{j=r,k=s}$$

must be square ($r = s \leqq \varkappa$) and nonsingular. But in this case the choice of the basis $(f_k)_{k=1}^r$ can be realized in such a way that

$$(e_j, f_k) = \delta_{jk} \quad (j,k = 1,\dots,r) . \tag{12.1}$$

Since Λ and Λ' are invariant for U, the operator U in L takes the form of a block matrix

$$U|L = \begin{bmatrix} V & 0 \\ 0 & W \end{bmatrix} ,$$

where the r-dimensional matrices $V = \|v_{jp}\|_{j,p=1}^r$ and $W = \|w_{kq}\|_{k,q=1}^r$ are defined for the bases $e_1,\dots,e_r$ and $f_1,\dots,f_r$, respectively, by the formulae

$$Ue_j = \sum_{p=1}^r v_{jp} e_p , \quad Uf_k = \sum_{q=1}^r w_{kq} f_q \quad (j,k = 1,\dots,r) . \tag{12.2}$$

From (12.1) and (12.2) we get that

$$\delta_{jk} = (e_j,f_k) = (Ue_j,Uf_k) = (\sum_{p=1}^{r} v_{jp}e_p, \sum_{q=1}^{r} w_{kq}f_q) = \sum_{p=1}^{r} v_{jp}\overline{w}_{kq}$$

$$(j,k = 1,\ldots,r) ,$$

i.e., $W = (V^*)^{-1}$.

Corollary 1. To each eigenvalue λ of a unitary operator U in $\Pi_\varkappa$ there corresponds an eigenvalue $\overline{\lambda}^{-1}$, and the elementary divisors of the operator referring to λ are in one-to-one correspondence with the elementary divisors referring to $\overline{\lambda}^{-1}$ in such a way that the orders of corresponding divisors are equal.

Corollary 1'. To each eigenvalue λ of a selfadjoint operator A there corresponds an eigenvalue $\overline{\lambda}$, so that the elementary divisors of the operator referring to λ are in one-to-one correspondence with the elementary divisors referring to $\overline{\lambda}$ in such a way that the orders of corresponding divisors are equal.

Corollary 2 (2'). In order that a number λ should be an eigenvalue of a unitary (selfadjoint) operator U (A) in $\Pi_\varkappa$ it is necessary and sufficient that the linear manifold

$$\{Ux - \lambda x \mid x \in \Pi_\varkappa\} \quad (\{Ax - \lambda x \mid x \in D(A)\} \text{ respectively})$$

should not be dense in $\Pi_\varkappa$. Hence $\sigma_r(U) = \emptyset$ ($\sigma_r(A) = \emptyset$).

Proof. Clearly, for the linear manifold $\{Ux - \lambda x\}$ not to be dense in $\Pi_\varkappa$ it is necessary and sufficient that there is a vector $x_o \neq \theta$ such that, for all $x \in \Pi_\varkappa$, $(Ux - \lambda x, x_o) = 0$. But from this we get the equation

$$(x, U^{-1}x_o - \overline{\lambda}x_o) = 0 \quad (x \in \Pi_\varkappa) ,$$

i.e., $U^{-1}x_o = \overline{\lambda}x_o$ and $Ux_o = \overline{\lambda}^{-1}x_o$. The last equation is possible if and only if λ is an eigenvalue of U (Corollary 1). For A the proof is analogous.

The corollaries just formulated are trivial for the case $|\lambda| = 1$ (for real λ, respectively). If $|\lambda| \neq 1$ (or for nonreal λ) they follow immediately from the fact that any root manifold S_λ lies in L, for which Theorem 12.2 (or 12.2') holds.

We may mention that the assertions of Corollaries 1 and 1' are concerned with operators which, essentially, act on finite-dimensional spaces of the type $\Pi_\varkappa$ (with nondegenerate scalar product) and for these the properties are well-known (see e.g., A.I. Mal'cev [31]).

12.4. On the basis of the preceding theorems and Theorem 3.2 for a given unitary operator U (selfadjoint operator A) we can represent the entire space $\Pi_\varkappa$ in the form of a direct sum $\Pi_\varkappa = L \oplus \Pi_{\varkappa'}$, where the subspace $\Pi_{\varkappa'}$ is invariant for U and such that all the eigenvalues of $U|\Pi_{\varkappa'}$ lie on the unit circle (invariant for A, $A|\Pi_{\varkappa'}$ is selfadjoint in $\Pi_{\varkappa'}$ and all its eigenvalues lie on the real axis respectively).

If the dimension of L is equal to 2r, then $\varkappa' = \varkappa - r$. Thus $0 \leqq \varkappa' \leqq \varkappa$. In particular, if $\varkappa' = 0$, then $\Pi_{\varkappa'} = \Pi_0$ is a negative unitary space, where the operator U (or A, respectively) acts as an ordinary unitary (selfadjoint) operator, the spectrum of which is completely determined by the corresponding spectral functions.

Note. If $\varkappa' = 0$, the space L is the direct sum of the pairs of root subspaces $S_{\lambda_j}(U)$, $S_{\bar{\lambda}_j^{-1}}(U)$ $(|\lambda_j| < 1;\ j = 1,\dots,p;\ p \geqq 1)$. We now form the subspace $\mathfrak{I}_0 = \text{l.s.}\ \{S_{\zeta_j}(U) \mid j = 1,\dots,p\}$, where for each j, $1 \leqq j \leqq p$, the corresponding ζ_j is either λ_j or $\bar{\lambda}_j^{-1}$. Evidently there are 2p such subspaces, and they are all neutral (see Theorem 7.1).

The space $\mathfrak{I}_0' = \mathfrak{I}_0^\perp$ is a nonpositive subspace containing $\mathfrak{I}_0$, and it is invariant under U.

Definition 12.2. A positive subspace M, invariant for U (A) and of maximal possible dimension is called an elliptic invariant subspace of U (A, respectively).

The elliptic invariant subspace M of U (A) belongs to $\Pi_{\varkappa'}$ and if $\dim M = s$, then $0 \leqq s \leqq \varkappa'$. We have $\Pi_{\varkappa'} = M \oplus \Pi_{\varkappa''}$, where $\varkappa'' = \varkappa' - s$, and the subspace $\Pi_{\varkappa''}$, invariant for U (A), contains no positive eigenvector of the operator considered.

If, now, $\varkappa'' = 0$, then the problem reduces once more to the study of an ordinary unitary (respectively selfadjoint) operator in the negative unitary space Π_0.

Thus, only the case when $\varkappa'' > 0$ remains to be considered.

Definition 12.3. The subspace $\Pi_{\varkappa''}$, i.e., the orthogonal complement of the direct sum $L \oplus M$ of the hyperbolic and elliptic subspaces (infinite dimensional if all $\Pi_\varkappa$ is infinite dimensional) will be called the parabolic subspace of the operator U (or A).

All the eigenvalues of the unitary operator U in $\Pi_{\varkappa''}$ clearly lie on the unit circle, and the selfadjoint operator A in $\Pi_{\varkappa''}$ has only real eigenvalues.

12.5. In the general case the whole space $\Pi_\varkappa$ is a direct sum of hyperbolic, elliptic and parabolic subspaces. Correspondingly each of the $\varkappa$-dimensional invariant subspaces $\mathfrak{J}$ and $\mathfrak{J}'$ decomposes into a direct sum of three orthogonal subspaces ("parts") with dimensions r, s, and $\varkappa''$.

It is clear that the "hyperbolic parts" of $\mathfrak{J}$ and $\mathfrak{J}'$ are Λ and Λ', respectively (see 12.2).

As for the "parts" of $\mathfrak{J}$ and $\mathfrak{J}'$, lying in the parabolic subspace $\Pi_{\varkappa''}$, they can coincide with each other, but can also be different (not orthogonal) subspaces. We give examples illustrating both cases.

As a first example we consider again the selfadjoint operator A in the finite-dimensional space $\Pi_\varkappa$ (of dimension $2\varkappa + 1$) defined by us in 8.5. It is easy to see that the parabolic subspace of this operator A coincides with $\Pi_\varkappa$. As for the $\varkappa$-dimensional invariant subspaces, they are here uniquely determined (i.e., $\mathfrak{J}$ coincides with $\mathfrak{J}'$). In fact, since $\Pi_\varkappa$ coincides with a cyclic subspace of A, corresponding to the elementary divisor $(\lambda - \alpha)^{2\varkappa+1}$, the $\varkappa$-dimensional invariant subspace must coincide with the linear span of the first $\varkappa$ vectors $e_1, \ldots, e_\varkappa$ of the "Jordan chain". This agrees completely with the fact that this linear span is neutral (the vector $e_{2\varkappa+1}$ is negative, $(e_{2\varkappa+1}, e_{2\varkappa+1}) = -1$).

As an example of the opposite sort we give the case of a linear operator U with a single eigenvalue ε ($|\varepsilon| = 1$) defined on the linear span L of the vectors $f_1, \ldots, f_{2\varkappa}$ by means of two "Jordan chains":

$$Uf_1 = \varepsilon f_1, \quad Uf_2 = f_1 + \varepsilon f_2, \; \ldots, \; Uf_\varkappa = f_{\varkappa-1} + \varepsilon f_\varkappa ,$$

$$Uf_{\varkappa+1} = \varepsilon f_{\varkappa+1}, \quad Uf_{\varkappa+2} = f_{\varkappa+1} + \varepsilon f_{\varkappa+2}, \; \ldots, \; Uf_{2\varkappa} = Uf_{2\varkappa-1} + \varepsilon f_{2\varkappa} .$$

If we introduce the metric by means of the Gram matrix

$$G = \| (f_j, f_k) \|_{j,k=1}^{2\varkappa} = \begin{bmatrix} 0 & 0 & \ldots & 0 & 1 \\ 0 & 0 & \ldots & 1 & 0 \\ \ldots & \ldots & \ldots & \ldots & \ldots \\ 0 & 1 & \ldots & 0 & 0 \\ 1 & 0 & \ldots & 0 & 0 \end{bmatrix} ,$$

then L becomes a space $\Pi_\varkappa$ of dimension $2\varkappa$, and U a unitary operator in this space. The parabolic subspace of U again coincides with the whole of $\Pi_\varkappa$, and the $\varkappa$-dimensional invariant nonnegative subspace here is not uniquely defined, since both $\mathfrak{J} = \text{l.s.}\{f_1,\dots,f_\varkappa\}$ and $\mathfrak{J}' = \text{l.s.}\{f_{\varkappa+1},\dots,f_{2\varkappa}\}$ have the **required properties** (in particular, both of them are neutral).

Each of these examples can be made into a "dual" example by means of the Cayley tranformation.

As the whole space $\Pi_\varkappa$ in the general case decomposes into a direct (orthogonal) sum of hyperbolic, elliptic and parabolic invariant subspaces of an operator U (or A), the spectrum of the respective operator is the set-theoretic sum of the spectrum of its "parts", acting in each of these subspaces.

From what has been said above it is clear that, as the hyperbolic and elliptic invariant subspaces are finite dimensional for both the unitary and the selfadjoint operators, the continuous spectrum $\sigma_c(U)$ ($\sigma_c(A)$) can appear only for the parabolic subspace $\Pi_{\varkappa''}$ (see Definition 12.3). If $\varkappa'' = 0$, then Π_0 is a negative (unitary) space in which U (or A) acts as a general unitary (selfadjoint, respectively) operator. The continuous spectrum of this operator (if such exists) exhausts the continuous spectrum of the corresponding operator in $\Pi_\varkappa$.

12.6. In this section we suppose again that $\dim \Pi_\varkappa > \varkappa$. Let A be a bounded selfadjoint operator in $\Pi_\varkappa$, $D(A) = \Pi_\varkappa$, for which there exists a real number υ such that

$$((A - \upsilon I)x,x) \geqq 0 \quad \text{for all } x \in \Pi_\varkappa \ . \qquad (12.3)$$

As an example, let T be a plus-operator in $\Pi_\varkappa$. Then, by Theorem 6.6 $A = T^*T$ has the property (12.3) with $\upsilon = \mu(T)$.

The relation (12.3) implies the existence of the two numbers

$$\upsilon_+(A) = \inf_{(x,x)=1} (Ax,x), \quad \upsilon_-(A) = \sup_{(x,x)=-1} (-(Ax,x)) \ ,$$

and we have $\upsilon_-(A) \leqq \upsilon \leqq \upsilon_+(A)$ if and only if υ satisfies (12.3). Evidently, if T satisfies (6.13), then $A = T^*T$ satisfies (12.3) and $\upsilon_\pm(A) = \mu_\pm(T)$.

In the following propositions $1^\circ - 5^\circ$ we always suppose that A satisfies (12.3). The proofs of the statements 2° and 4° in 6.5 give the following proposition:

1°. If $\upsilon_-(A) < \upsilon_+(A)$, then for each υ, $\upsilon_-(A) < \upsilon < \upsilon_+(A)$, there exists a $\delta(\upsilon) > 0$ such that

$$((A - \upsilon I)x,x) \geqq \delta(\upsilon)\|x\|^2 \quad (x \in \Pi_\varkappa) ,$$

and the interval $(\upsilon_-(A),\upsilon_+(A))$ belongs to $\rho(A)$.

2°. Let φ be an eigenvector of A: $A\varphi = \lambda\varphi$. Then:

(1) If $(\varphi,\varphi) = 0$, then $\lambda = \upsilon_-(A) = \upsilon_+(A)$;

(2) if $(\varphi,\varphi) > 0$, then $\lambda \geqq \upsilon_+(A)$;

(3) if $(\varphi,\varphi) < 0$, then $\lambda \leqq \upsilon_-(A)$.

If $(\varphi,\varphi) \neq 0$, i.e. in cases (2) and (3), there is no solution ψ of the equation $(A - \lambda I)\psi = \varphi$; that is, φ does not have associated vectors.

Indeed, $A\varphi = \lambda\varphi$ implies

$$0 \leqq ((A - \upsilon I)\varphi,\varphi) = (\lambda - \upsilon)(\varphi,\varphi); \tag{12.4}$$

hence from $(\varphi,\varphi) = 0$ we have $((A - \upsilon I)\varphi,\varphi) = 0$, and (12.3) implies $(A - \upsilon I)\varphi = \theta$. It follows that $\lambda = \upsilon_-(A) = \upsilon_+(A)$. In cases (2) and (3) the statements follow from (12.4). The last statement in 2° is a general fact for selfadjoint operators in $\Pi_\varkappa$: $A\psi - \lambda\psi = \varphi$ implies $(A\psi - \lambda\psi,\varphi) = (\varphi,\varphi) = (\psi,A\varphi - \lambda\varphi) = 0$, a contradiction if $(\varphi,\varphi) \neq 0$.

From Theorem 6.4 and 2° we have now immediately the following two statements:

3°. $\sigma(A)$ is real.

4°. If $\lambda \in \sigma_p(A)$, $\lambda > \upsilon_-(A)$ (or $\lambda < \upsilon_+(A)$), then $L_\lambda(A) = S_\lambda(A)$ and this subspace is positive (negative resp.).

5°. If $\upsilon_-(A) = \upsilon_+(A)$, then $\lambda_0 := \upsilon_\pm(A) \in \sigma_p(A)$ and $(A - \lambda_0 I)^2 S_{\lambda_0}(A) = \{\theta\}$.

We shall not prove the claim $\lambda_0 \in \sigma_p(A)$ here, it will follow from Theorem 12.3, 2) below. Let n be the smallest integer such that $(A - \lambda_0 I)^n S_{\lambda_0}(A) = \{\theta\}$. Then $n \leqq 2\varkappa + 1$ (see 8.5). Assume $n \geqq 3$. Then $2n - 3 \geqq n$ and for arbitrary $y \in \Pi_\varkappa$, $x \in S_{\lambda_0}(A)$, from (12.3) it follows

$$|((A - \lambda_0 I)(A - \lambda_0 I)^{n-2}x,y)|^2 \leqq$$
$$\leqq ((A - \lambda_0 I)(A - \lambda_0 I)^{n-2}x,(A - \lambda_0 I)^{n-2}x)((A - \lambda_0 I)y,y) =$$
$$= ((A - \lambda_0 I)^{2n-3}x,x)((A - \lambda_0 I)y,y) = 0 ,$$

that is $(A - \lambda_0 I)^{n-1} x = \theta$, a contradiction to the minimality of n.

According to Theorem 12.1' and its Note, the selfadjoint operator A has a $\varkappa$-dimensional nonnegative (maximal nonpositive) invariant subspace $\mathfrak{J}_+$ ($\mathfrak{J}_-$, respectively). Some properties of these subspaces will be established in the following theorem.

Theorem 12.3. Suppose the selfadjoint operator A satisfies (12.3).
1) Let $\upsilon_-(A) < \upsilon_+(A)$. Then A has a unique $\varkappa$-dimensional nonnegative (maximal nonpositive) invariant subspace $\mathfrak{J}_+$ ($\mathfrak{J}_-$ respectively). Both subspaces $\mathfrak{J}_+$ and $\mathfrak{J}_-$ are definite, $\Pi_\varkappa = \mathfrak{J}_+ \oplus \mathfrak{J}_-$ and

$$\max \sigma(A|\mathfrak{J}_-) = \upsilon_-(A) < \upsilon_+(A) = \min \sigma(A|\mathfrak{J}_+) .$$

2) Let $\upsilon_-(A) = \upsilon_+(A) =: \lambda_0$. Then an arbitrary $\varkappa$-dimensional nonnegative subspace $\mathfrak{J}_+$ (maximal nonpositive subspace $\mathfrak{J}_-$) which is invariant under A satisfies

$$\lambda_0 = \min \sigma(A|\mathfrak{J}_+) \quad (= \max \sigma(A|\mathfrak{J}_-)) .$$

Here $\lambda_0 \in \sigma_p(A)$ and there exist nonnegative eigenvectors corresponding to the eigenvalue λ_0. Moreover, $\mathfrak{J}_+ \cap \mathfrak{J}_+^{\perp} \subset S_{\lambda_0}(A)$. If $\mathfrak{J}_+$ does not contain neutral vectors ($\mathfrak{J}_+ \cap \mathfrak{J}_+^{\perp} = \{\theta\}$), then $\mathfrak{J}_+$ and $\mathfrak{J}_+^{\perp}$ are definite invariant subspaces and $\Pi_\varkappa = \mathfrak{J}_+ \oplus \mathfrak{J}_+^{\perp}$.

Proof. Let $\mathfrak{J}_+$ ($\mathfrak{J}_-$) be an arbitrary $\varkappa$-dimensional nonnegative (maximal nonpositive, respectively) subspace invariant under A. Then

$$\sigma(A|\mathfrak{J}_+) \cup \sigma(A|\mathfrak{J}_-) \subset \sigma(A) , \tag{12.5}$$

as the set $\rho(A)$ is connected. Moreover,

$$\sigma(A|\mathfrak{J}_+) \geqq \upsilon_+(A), \quad \sigma(A|\mathfrak{J}_-) \leqq \upsilon_-(A) . \tag{12.6}$$

Here the first relation is an immediate consequence of 2° and 3°. To prove the second relation let $\lambda \in \sigma(A|\mathfrak{J}_-)$. Then there exists a sequence $(x_n) \subset \mathfrak{J}_-$, $\|x_n\| = 1$, such that $(A - \lambda I)x_n \to \theta$ $(n \to \infty)$. As $|(x_n,x_n)| \leqq 1$, we may assume that $c := \lim_{n\to\infty}(x_n,x_n)$ $(\leqq 0)$ exists (otherwise we can consider a suitable subsequence). Then

$$c(\upsilon_-(A) - \lambda) = \lim_{n\to\infty}(\upsilon_-(A) - \lambda)(x_n,x_n) \leqq \lim_{n\to\infty}((A - \lambda I)x_n,x_n) = 0.$$

If $c < 0$, it follows that $\lambda \leqq \upsilon_-(A)$. Now let $c = 0$. As $Ax_n \in \mathfrak{J}_-$, we find that $(Ax_n,x_n) \to 0$ $(n \to \infty)$. Moreover, for arbitrary $\upsilon \in [\upsilon_-(A),\upsilon_+(A)]$ we have

$$|((A - \upsilon I)x_n, y)|^2 \leqq (A - \upsilon I)x_n, x_n)\|A - \upsilon I\|\|y\|^2 \quad (y \in \Pi_\varkappa) .$$

This implies $(A - \upsilon I)x_n \to \theta$. $(n \to \infty)$, and therefore $\upsilon = \upsilon_+(A) = \upsilon_-(A) = \lambda$.

To prove part 1) of Theorem 12.3, suppose $\upsilon_-(A) < \upsilon_+(A)$. Then, according to 4°, all the eigenspaces $L_\lambda(A)$, $\lambda \in \sigma_p(A)$, $\lambda \geqq \upsilon_+(A)$, are positive. Hence their linear span is the unique $\varkappa$-dimensional nonnegative (even positive) invariant subspace $\mathfrak{J}_+$ of A. If a maximal nonpositive invariant subspace $\mathfrak{J}_-$ is given, $\mathfrak{J}_+ = \mathfrak{J}_-^{\perp}$ can be considered, and the uniqueness and definiteness follow. From (12.6) we have

$$\rho_- := \max \sigma(A|\mathfrak{J}_-) \leqq \upsilon_-(A) < \upsilon_+(A) \leqq \min \sigma(A|\mathfrak{J}_+) =: \rho_+ .$$

In order to show that instead of $\leqq$ we in fact have equalities, we use the fact that $\mathfrak{J}_+$ ($\mathfrak{J}_-$) is a Hilbert space with respect to the scalar product (x,y) $(-(x,y))$ and that A is selfadjoint in this Hilbert space. So for $x = x_+ + x_-$, $x_\pm \in \mathfrak{J}_\pm$, we get

$$(Ax,x) = (Ax_+,x_+) + (Ax_-,x_-) \geqq \rho_+(x_+,x_+) + \rho_-(x_-,x_-) \geqq$$

$$\geqq \rho_\pm((x_+,x_+) + (x_-,x_-)) = \rho_\pm(x,x) ,$$

and it follows that $\rho_\pm \in [\upsilon_-(A),\upsilon_+(A)]$, or $\rho_\pm = \upsilon_\pm(A)$.

Suppose now $\upsilon_-(A) = \upsilon_+(A) =: \lambda_o$. Then (12.6) gives $\sigma(A|\mathfrak{J}_+) \geqq \lambda_o$. Assume that for some $\mathfrak{J}_+$ the sign > holds. Then, according to 4°, the $\varkappa$-dimensional nonnegative invariant subspace $\mathfrak{J}_+$ of A is positive and uniquely determined as the linear span of all $L_\lambda(A)$, $\lambda > \lambda_o$, and (12.7), applied to $\mathfrak{J}_+$ and $\mathfrak{J}_- = \mathfrak{J}_+^{\perp}$, gives $\upsilon_+(A) > \lambda_o$, a contradiction. From (12.6) we also have $\sigma(A|\mathfrak{J}_-) \leqq \lambda_o$. Assume that the sign < holds. Then the eigenspace $L_{\lambda_o}(A)$ is positive (otherwise $\mathfrak{J}_-$ would not be a maximal nonpositive subspace), the $\varkappa$-dimensional nonnegative subspace $\mathfrak{J}_+$ is uniquely determined and from (12.7) we get $\upsilon_-(A) < \lambda_o$, a contradiction.

If $L := \mathfrak{J}_+ \cap \mathfrak{J}_+^{\perp}$, then $AL \subset L$ and for each eigenvalue λ of the restriction of A to L according to 2° we have $\lambda = \upsilon_-(A) = \upsilon_+(A)$, hence $L \subset S_{\lambda_o}(A)$. The remaining statements of 2) are now clear and the theorem is proved.

Note. We mention without proof, that the relations (12.5) and (12.6) can be sharpened in the following way. If $\mathfrak{J}_+$ ($\mathfrak{J}_-$) is an arbitrary $\varkappa$-dimensional nonnegative (maximal nonpositive, respectively) invariant subspace of $\mathcal{K}$ satisfying (12.3), then

$$\sigma(A) = \sigma(A|\mathfrak{J}_+) \cup \sigma(A|\mathfrak{J}_-),$$

$$\sigma(A|\mathfrak{J}_-) = \sigma(A) \cap (-\infty, \upsilon_-(A)],\ \sigma(A|\mathfrak{J}_+) = \sigma(A) \cap [\upsilon_+(A), \infty) .$$

According to the statement 1) of Theorem 12.3, if the operator A satisfies (12.3) and $\upsilon_-(A) < \upsilon_+(A)$, then A is the orthogonal sum $A = A_+ \oplus A_-$ of the nonnegative operators A_+, A_- in the Hilbert spaces $\mathfrak{J}_+$ (with scalar product (x,y)) and $\mathfrak{J}_-$ (with scalar product - (x,y)) respectively. Their spectra are divided by the gap $(\upsilon_-(A), \upsilon_+(A))$, and a spectral decomposition for A can be obtained in an obvious way:

$$A = \int_{-\|A\|}^{\upsilon_-(A)} \lambda \, dE(\lambda) + \sum_{j=1}^{\ell} \rho_j P_j ,$$

where $\upsilon_+(A) = \rho_1 < \rho_2 < \dots < \rho_\ell$, P_j $(j = 1,2,\dots,\ell)$ are orthogonal projectors in $\Pi_\varkappa$ with range in $\mathfrak{J}_+$, $\sum_{j=1}^{\ell} \dim P_j = \varkappa$ and E is the spectral function of A_-, extended to all of $\Pi_\varkappa$ by $E(\Delta)\mathfrak{J}_+ = \{\theta\}$.

H. Langer [28d] and, by another method, M.G. Kreĭn [26b] obtained a spectral decomposition of nonnegative operators in a space Π with regular indefinite metric, and, in particular, in $\Pi_\varkappa$. Using this spectral decomposition it is easy to prove the results of this section.

12.7. As was mentioned at the beginning of 12.6, the statements 2° - 5° as well as Theorem 12.3 immediately apply to $A = T^*T$, where T is a plus-operator, and we have $\upsilon_\pm(A) = \mu_\pm(T)$. If the plus-operator is even strict, the following theorem holds.

Theorem 12.4. Let T be a strict plus-operator in $\Pi_\varkappa$. Then:

1) $\mu_\pm(T^*T) = \mu_\pm(TT^*) = \mu_\pm(T)^2$;

2) $\sigma(T^*T) \subset [0,\infty)$.

Proof. As T and $T^*T = A$ are strict plus-operators (see Corollary 1 of Lemma 11.3') from 6.5, 3° we have $\mu_-(T) \geqq 0$ and $\mu_-(T^*T) \geqq 0$. If $\mu_-(T) \leqq \mu \leqq \mu_+(T)$, then

$$(T^*Tx, T^*Tx) \geqq \mu(Tx, Tx) \geqq \mu^2(x,x) \qquad (x \in \Pi_\varkappa) ,$$

hence $\mu_+(T^*T) \geqq \mu_+(T)^2$, $\mu_-(T^*T) \leqq \mu_-(T)^2$. Assume that e.g. in the second relation the sign < holds. Then, by 1°, for $(0 \leqq)$ $\mu_-(T^*T) < \upsilon < \mu_-(T)^2$ we have

$$(T^*Tx, T^*Tx) \geqq \upsilon(x,x) + \delta(\upsilon)\|x\|^2 \quad (x \in \Pi_\varkappa)$$

with some $\delta(\upsilon) > 0$. On the other hand, by Theorem 12.3, $\mu_-(T) = \upsilon_-(A)$ belongs to $\sigma(A|\mathfrak{J}_-)$, therefore there exists a sequence (x_n), $\|x_n\| = 1$, $(x_n, x_n) \leqq 0$ such that $(A - \mu_-(T)I)x_n \to \theta$ $(n \to \infty)$. Then

$$\delta(\upsilon)\|x_n\|^2 \leqq (Ax_n, Ax_n) - \upsilon(x_n, x_n) \leqq (Ax_n, x_n) - \mu_-(T)^2(x_n, x_n) ,$$

and the right hand side tends to zero if $n \to \infty$, a contradiction, that is, $\mu_-(T^*T) = \mu_-(T)^2$.

To prove statement 2) assume $\lambda := \inf \sigma(T^*T) < 0$. Then $\lambda < 0 \leqq \upsilon_-(A) = \mu_-(T)$, hence there exists a sequence (x_n), $\|x_n\| = 1$, such that

$$(A - \lambda I)x_n \to \theta \quad (n \to \infty) . \tag{12.8}$$

As $|(x_n, x_n)| \leqq 1$, we may again suppose that $\lim_{n\to\infty}(x_n, x_n) = c$ exists. If $c > 0$, then for $\upsilon \in [\upsilon_-(A), \upsilon_+(A)]$ we get $(\lambda - \upsilon) = \frac{1}{c} \lim_{n\to\infty}((A - \upsilon I)x_n, x_n) \geqq 0$, that is $\lambda \geqq \upsilon$, a contradiction. If $c = 0$, then (12.8) implies $(Ax_n, x_n) \to 0$ $(n \to \infty)$ and as in the proof of Theorem 3.9 we get

$$((A - \upsilon_-(A)I)x_n, x_n) \to 0, \quad (A - \upsilon_-(A)I)x_n \to \theta, \quad (n \to \infty) ,$$

hence $\lambda = \upsilon_-(A) = \mu_-(T) \geqq 0$, a contradiction. If $c < 0$ then we have from Corollary 2 of Lemma 11.3'

$$0 \leqq (A(A - \upsilon_-(A)I)x_n, x_n) \to \lambda(\lambda - \upsilon_-(A))c < 0 ,$$

again a contradiction. The proof of Theorem 12.4 is complete.

In the same way as we have shown some properties of the selfadjoint operator T^*T if T is a plus-operator (or an expansive operator), similar assertions can be proved for T^*T if T is a contractive operator. The analogue of Theorem 12.4, 2) is e.g. as follows:

(*) If T is a doubly contractive operator in $\Pi_\varkappa$, then $\sigma(T^*T) \subset [0, \infty)$.

In a finite-dimensional space each contractive operator is doubly contractive. In this case the statement (*) is contained in the considerations of V.P. Potapov in [36]. Ju.P. Ginsburg [16a] proved this assertion for doubly contractive operators in an arbitrary infinite-dimensional space Π with a regular indefinite metric; of course, this implies also 2) of Theorem 12.4.

S U P P L E M E N T

THE GENERAL FORM OF UNITARY AND SEMI-UNITARY OPERATORS IN A SPACE $\Pi_\varkappa$

1. Let U be a unitary or semi-unitary operator in $\Pi_\varkappa$. We choose a decomposition (1.1) of $\Pi_\varkappa$ and consider the corresponding matrix representation of U:

$$U = \begin{bmatrix} A & B \\ C & D \end{bmatrix} . \tag{S. 1}$$

For simplicity, we denote the H-adjoint operators of the components B and C by B^* and C^* (instead of $B^{[*]}$ and $C^{[*]}$ in 5.2). It follows easily that

$$U^* = \begin{bmatrix} A & -C^* \\ -B^* & D \end{bmatrix} .$$

Hence the relation $U^*U = I$, written in its components, gives

$$A^*A - C^*C = I_+,\ D^*D - B^*B = I_-,\ B^*A - D^*C = 0 , \tag{S. 2}$$

where I_+ (I_-) is the identity operator in Π_+ (Π_-).

The H-Hermitian operators A^*A, B^*B, C^*C and D^*D are obviously non-negative (in the sense of the nonnegativeness of the corresponding Hermitian forms), and the relations (S. 2) show that A^*A and D^*D are strictly positive, or symbolically,

$$A^*A \geqq I_+ > 0,\quad D^*D \geqq I_- > 0 . \tag{S. 3}$$

The first of the relations (S. 3) means that $[A^*Ax_+,x_+] > 0$ $(x_+ \neq \theta)$, i.e. $Ax_+ \neq \theta$ if $x_+ \neq \theta$; and this implies that A^{-1} exists. In the same way $Dx_- \neq \theta$ for $x_- \neq \theta$.

2. If U is in fact unitary, then we also have $UU^* = I$; that is,

$$AA^* - BB^* = I_+,\ DD^* - CC^* = I_-,\ AC^* - BD^* = 0 .$$

It follows $DD^* \geqq I_-$, and because we know that D is ivertible, it maps Π_- one-to-one onto Π_-; that is, D^{-1} exists.

Now it is not difficult to find the general form of a unitary operator in $\Pi_\varkappa$.

In fact, we define an operator Γ by

$$CA^{-1} = \Gamma, \quad \text{then } A^{*-1}C^* = \Gamma^* . \tag{S. 4}$$

From (S. 1) and (S. 3) we obtain

$$A^*(I_+ - \Gamma^*\Gamma)A = I_+, \quad I_+ - \Gamma^*\Gamma = A^{*-1}A^{-1} = (AA^*)^{-1} . \tag{S. 5}$$

From this it follows that $I_+ - \Gamma^*\Gamma$ is a strictly positive operator:

$$I_+ - \Gamma^*\Gamma > 0, \; \Gamma^*\Gamma < I_+ .$$

Now we define an operator U_+ in Π_+ by the formula

$$U_+ = (I_+ - \Gamma^*\Gamma)^{1/2}A . \tag{S. 6}$$

Then $U_+^* = A^*(I_+ - \Gamma^*\Gamma)^{1/2}$, and, from (S. 5), $U_+^*U_+ = I_+$, whence it follows that U_+ is a unitary operator in the (finite dimensional) space Π_+.

From (S. 6) and (S. 4) we now find that

$$A = (I_+ - \Gamma^*\Gamma)^{-1/2}U_+, \; C = \Gamma(I_+ - \Gamma^*\Gamma)^{-1/2}U_+ . \tag{S. 7}$$

Returning to the last relation of (S. 2) we notice that

$$A^*B = C^*D, \quad A^*BD^{-1} = C^*, \quad BD^{-1} = A^{*-1}C^* = (CA^{-1})^* = \Gamma^* ,$$

i.e.,

$$BD^{-1} = \Gamma^*, \quad D^{*-1}B^* = \Gamma . \tag{S. 8}$$

The second formula of (S. 2) gives $D^*(I_- - \Gamma\Gamma^*)D = I_-$, where $I_- - \Gamma\Gamma^* = D^{*-1}D^{-1} = (DD^*)^{-1}$, which is again a strictly positive operator, i.e. $\Gamma\Gamma^* < I_-$. The operator $I_- - \Gamma\Gamma^*$ maps Π_- onto Π_- (for the operator DD^* acts in this way). The same applies, clearly, to the operators $(I_- - \Gamma\Gamma^*)^{1/2}$ and $(I_- - \Gamma\Gamma^*)^{-1/2}$.

Writing

$$U_- = (I_- - \Gamma\Gamma^*)^{1/2}D, \quad \text{whence } U_-^* = D^*(I_- - \Gamma\Gamma^*)^{1/2} , \tag{S. 9}$$

we find that $U_-^*U_- = I_-$.

In addition, the first of the relations (S. 9) shows that $D\Pi_- = \Pi_-$ implies that $U_-\Pi_- = \Pi_-$, and conversely. Thus U_- is a unitary operator in Π_-.

The formulae (S. 9) and (S. 8) now give

$$D = (I - \Gamma\Gamma^*)^{-1/2}U_-, \quad B = \Gamma^*(I - \Gamma\Gamma^*)^{-1/2}U_- . \qquad (S. 10)$$

Theorem. In order that the linear operator U in $\Pi_\varkappa$ should be unitary, it is necessary and sufficient that it is given by (S. 1) with

$$A = (I_+ - \Gamma^*\Gamma)^{-1/2}U_+, \quad B = \Gamma^*(I_- - \Gamma\Gamma^*)^{-1/2}U_- ,$$
$$C = \Gamma(I_+ - \Gamma^*\Gamma)^{-1/2}U_+, \quad D = (I_- - \Gamma\Gamma^*)^{-1/2}U_- , \qquad (S. 11)$$

where U_+ and U_- are unitary operators acting in Π_+ and Π_-, respectively, and Γ is a bounded linear operator from Π_+ to Π_- satisfying the conditions

$$\Gamma^*\Gamma < I_+, \ \Gamma\Gamma^* < I_-, \ (I - \Gamma\Gamma^*)\Pi_- = \Pi_- .$$

Proof. The necessity has been established above. To prove the sufficiency one only has to verify that the operator U given by (S. 1) and (S. 11) satisfies $U^*U = I$ and $UU^* = I$. The first relation follows easily by multiplication of the matrices of U and U^*; to prove the second relation in the same way we have to make use of the well-known relation $\Gamma(I - \Gamma^*\Gamma)^{-1/2} = (I - \Gamma\Gamma^*)^{-1}\Gamma$.

Note 1. Incidentally we have obtained the general form of semi-unitary operators in $\Pi_\varkappa$. In fact, returning to the first of formulae (S. 10), we see that if D takes Π_- into a proper part of Π_-, then U has the same property, i.e., is not unitary but semi-unitary. Conversely, if in (S. 11) we require of the operator U_- in Π_- only that it be semi-unitary (i.e., only that $U_-^*U_- = I_-$), then the operator U constructed according to (S. 11) will be a semi-unitary operator in $\Pi_\varkappa$.

Note 2. Consider the fractional transformation $K \to \hat{K}$ in Note 1 after Lemma 10.3 for the unitary operator U instead of T. Then, with (S. 11), it takes the form

$$\hat{K} = - U_-(I_- - \tau\tau^*)^{-1/2}(\tau - K)(I_+ - \tau^*K)^{-1}(I_+ - \tau^*\tau)^{1/2}U_+, \qquad (S. 12)$$

where $\tau = - U_-^*\Gamma U_+$, i.e., $\tau^*\tau < I_+$ and $\tau\tau^* < I_-$.

It is known (see, e.g., V.P. Potapov [36]) that in the case of finite square matrices, (S. 12) gives the general form of the one-to-one analytic transformations of the "matrix unit circle" into itself. R.S. Phillips [34b] generalized this result to the infinite-dimensional case.

BIBLIOGRAPHY

[1] **V.M. Adamjan, D.Z. Arov, M.G. Krein** /В.М.Адамян, Д.З.Аров, М.Г.Крейн/
Аналитические свойства пар Шмидта ганкелева оператора и обобщённая задача Шура-Такаги, Мат.сб. 86 /1971/, 34-75.

[2] **N.I. Ahiezer and I.M. Glazman** /Н.И.Ахиезер, И.М.Глазман/
Теория линейных операторов в гильбертовом пространстве, Москва, Наука, 1966.
German transl.: Akademie-Verlag, Berlin, 1968.
Engl. transl. of 1-st ed.: Ungar, New York, 1962.

[3] **T.Ja. Azizov** /Т.Я.Азизов/
a) Инвариантные подпространства и критерии полноты системы корневых векторов $\mathcal{J}$-диссипативных операторов в пространстве Понтрягина $\Pi_\varkappa$, Докл. АН СССР, 200, № 5 /1971/, 1015-1017.
Engl. transl.: Soviet Math. Dokl. 12 (1971), 1513 - 1516.
b) Диссипативные операторы в гильбертовом пространстве с индефинитной метрикой, Изв. АН СССР Сер.мат. 37, № 3 /1973/, 639-662.
Engl. transl.: Math. USSR Izv. 7 (1973), 639 - 660.

[4] **T.Ja. Azizov and I.S. Iohvidov** /Т.Я.Азизов, И.С.Иохвидов/
a) Линейные операторы в гильбертовых пространствах с G-метрикой, Успехи мат. наук 26, № 4 /1971/, 43-92.
Engl. transl.: Russian Math. Surveys 24, 4 (1971), 45 - 97.
b) Критерий полноты и базисности корневых векторов вполне непрерывного $\mathcal{J}$-самосопряжённого оператора в пространстве Понтрягина $\Pi_\varkappa$, Мат. исследования 6, вып. I /1971/, 158-161.
c) Линейные операторы в пространствах с индефинитной метрикой и их приложения, Итоги науки и техники, серия матем. анализ, том 17 /1979/, 113-205.

[5] **T.Ja. Azizov and E.I. Iohvidov** /Т.Я.Азизов, Е.И.Иохвидов/
Об инвариантных подпространствах максимальных $\mathcal{J}$-диссипативных операторов, Мат. заметки 12 /1972/, 747-754.
Engl. transl.: Math. Notes 12 (1972), 886 - 889.

[6] **T.Ja. Azizov and E.B. Usvjacova** /Т.Я.Азизов, Е.Б.Усвяцова/
Существование специальных неотрицательных подпространств $\mathcal{J}$-диссипативных операторов, Сиб. мат. ж. 18 /1977/, 216-218.
Engl. transl.: Sibirian Math. J. 18 (1977), 158 - 159.

[7] **T.Ja. Azizov, S.A. Horošavin** /Т.Я.Азизов, С.А.Хорошавин/
Об инвариантных подпространствах операторов, действующих в пространстве с индефинитной метрикой, Функц. анализ и его прилож. I4,4 /I980/, I - 7.

[8] **S. Banach**
Theorie des operations lineaires, PWN, Warsaw, 1932, reprint: Chelsea, New York, 1955.

[9] **J. Bognár**
Indefinite inner product spaces, Springer-Verlag, Berlin and New York, 1974.

[10] **M.L. Brodskii** /М.Л.Бродский/
О свойствах оператора, отображающего в себя неотрицательную часть пространства с индефинитной метрикой, Успехи мат. наук I4, № I /I959/, I47-I52.

[11] **Ju.L. Daleckii and M.G. Krein** /Ю.Л.Далецкий, М.Г.Крейн/
Устойчивость решений дифференциальных уравнений в банаховом пространстве, Наука, Москва, I970.
Engl. transl.: Amer. Math. Soc., Providence, R. I., 1974.

[12] **L.L. Dines**
On the mapping of quadratic forms, Bull. Amer. Math. Soc. 47 (1941), 494 - 498.

[13] **N. Dunford and J.T. Schwartz**
Linear operators. Vol. I: General theory, Interscience New York, 1958.

[14] **P. Finsler**
Über das Vorkommen definiter und semidefiniter Formen in Scharen quadratischer Formen, Comm. Math. Helvetici 9 (1937), 188 - 192.

[15] **F.R. Gantmaher and M.G. Krein** /Ф.Р.Гантмахер, М.Г.Крейн/
Осцилляционные матрицы и ядра и малые колебания механических систем, ГИТТЛ, Москва, I950.
German transl.: Akademie-Verlag, Berlin, 1960.

[16] **Ju. P. Ginzburg** /Ю.П.Гинзбург/
a) О $\mathfrak{J}$-нерастягивающих операторах в гильбертовом пространстве, Научн. зап. физ.-мат. фак. Одесск. гос. пед. ин-та 22, № I /I958/, I3-20.
b) О подпространствах гильбертова пространства с индефинитной метрикой, Научн. зап. кафедр мат., физ. и естествозн. Одесск. гос. пед. ни-та 25, № 2 /I96I/, 3 - 9.

[17] **I.C. Gohberg, M.G. Krein** /И.Ц.Гохберг, М.Г.Крейн/

a) Основные положения о дефектных числах, корневых числах и индексах линейных операторов, Успехи мат. наук I2, № 2 /I957/, 43 - II8.
Engl. transl.: Amer. Math. Soc. Transl. (2) 13 (1960), 185 - 264.

b) Введение в теорию линейных несамосопряжённых операторов в гильбертовом пространстве, Наука, Москва, I965.
Engl. transl.: Amer. Math. Soc., Providence, R. I., 1969.

[18] **I.S. Iohvidov** /И.С.Иохвидов/

a) Унитарные операторы в пространстве с индефинитной метрикой, Зап. НИИ матем. и механ. ХГУ и Харьк. матем. о-ва /4/ 2I /I949/, 79-86.

b) Унитарные и самосопряжённые операторы в пространстве с индефинитной метрикой, Канд. дисс. Одесса, I950.

c) О спектрах эрмитовых и унитарных операторов в пространстве с индефинитной метрикой, Докл. АН СССР 7I, № 2 /I950/, 225 - 228.

d) К теории неопределённых теплицевых форм, Докл. АН СССР I0I, № 2 /I955/, 2I3 - 2I6.

e) О пространствах финитных последовательностей с невырожденной эрмитовой формой, Тр. НИИ мат. Воронеж. ун-та 20 /I975/, 30-38.

f) О банаховых пространствах с $\mathfrak{J}$-метрикой и некоторых классах линейниых операторов в этих пространствах, Изв. АН Молд. ССР сер. физ.-тех. и мат. н. I /I968/, 60 - 80.

g) Дробно-линейные преобразования $\mathfrak{J}$-нерастягивающих операторов, Докл. АН Арм. ССР 42, № I /I966/, 3 - 8.

[19] **I.S. Iohvidov and M.G. Krein** /И.С.Иохвидов, М.Г.Крейн/

a) Спектральная теория операторов в пространствах с индефинитной метрикой. I, Тр. Моск. мат. об-ва 5 /I956/, 367 - 432.
Engl. transl.: Amer. Math. Soc. Transl. (2) 13 (1960), 105 - 175.

b) Спектральная теория операторов в пространствах с индефинитной метрикой. II, Тр. Моск. мат. об-ва 8 /I959/, 4I3 - 496.
Engl. transl.: Amer. Math. Soc. Transl. (2) 34 (1963), 283 - 373.

[20] **T. Kato**
Perturbation theory for linear operators, Springer-Verlag, Berlin and New York, 1966.

[21] **M.G. Krein** /М.Г.Крейн/

a) Об одном применении принципа неподвижной точки в теории линейных преобразований пространств с индефинитной метрикой, Успехи мат. наук 5, № 2 /1950/, 180 - 190.
Engl. transl.: Amer. Math. Soc. Transl. (2) 1 (1955), 27 - 35.

b) О применении одного алгебраического предложения в теории матриц монодромии, Успехи мат. наук 6, № 1 /1951/, 171 - 177.

c) Основные положения теории λ-зон устойчивости канонической системы линейных дифференциальных уравнений с периодическими коэффициентами, Сб. памяти А.А.Андронова, Изд. АН СССР, 1955, 413 - 498.

d) Винтовые линии в пространстве Лобачевского бесконечного числа измерений и лоренцовы преобразования, Успехи мат. наук 3, № 3 /1948/, 158-160.

e) Об интегральном представлении непрерывной эрмитово-индефинитной функции с конечным числом отрицательных квадратов, Докл. АН СССР 125, № 1 /1959/, 31 - 34.

f) Об одном новом принципе применения метода неподвижной точки в теории операторов в пространстве с индефинитной метрикой, Докл. АН СССР 154, № 5 /1964/, 1023 - 1026.
Engl. transl.: Soviet Math. Dokl. 5 (1964), 224 - 227.

g) Лекции по теории устойчивости решений дифференциальных уравнений в банаховом пространстве, АН УССР, Институт математики, Киев, 1964.

[22] **M.G. Krein and M.A. Krasnosel'skiĭ** /М.Г.Крейн,М.А.Красносельский/
Устойчивость индекса неограниченного оператора, Мат. сб. 30 /1952/, 219 - 224.

[23] **M.G. Kreĭn, M.A. Krasnosel'skiĭ and D.P. Mil'man** /М.Г.Крейн, М.А.Красносельский, Д.П.Мильман/
О дефектных числах линейных операторов в банаховом пространстве и о некоторых геометрических вопросах, Сб. трудов Института матем. АН УССР 11 /1948/, 97 - 113.

[24] **M.G. Kreĭn and H. Langer**

a) О спектральной функции самосопряжённого оператора в пространстве с индефинитной метрикой, Докл. АН СССР 152, № 1 /1963/, 39 - 42.
Engl. transl.: Soviet Math. Dokl. 4 (1963), 1236 - 1239.

b) **Über die verallgemeinerten Resolventen und die charakteristische Funktion eines isometrischen Operators im Raume $\Pi_{\varkappa}$, Hilbert Space Operators and Operator Algebras (Proc. Internat. Conf., Tihany, 1970), Colloqu. Math. Soc. János Bolyai, no. 5, North- Holland, Amsterdam, 1972, 353 - 399.**

c) Über die Q-Funktion eines π-hermiteschen Operators im Raume $\Pi_\varkappa$, Acta Sci. Math. (Szeged) 34 (1973), 191 - 230.

d) О дефектных подпространствах и обобщённых резольвентах эрмитова оператора в пространстве $\Pi_\varkappa$, Функц. анализ и его прилож. 5,2 /1971/, 59 - 71; 5,3 /1971/, 54 - 69.
Engl. transl.: Functional Anal. Appl. 5 (1971), 136 - 146; 217 - 228.

e) Über einige Fortsetzungsprobleme, die eng mit der Theorie hermitescher Operatoren im Raume $\Pi_\varkappa$ zusammenhängen.
Teil I: Einige Funktionenklassen und ihre Darstellungen, Math. Nachr. 77 (1977), 187 - 236.

f) Über einige Fortsetzungsprobleme, die eng mit der Theorie hermitescher Operatoren im Raume $\Pi_\varkappa$ zusammenhängen.
Teil II: Verallgemeinerte Resolventen, u-Resolventen und ganze Operatoren, J. Functional Analysis 30 (1978), 390 - 447.

[25] M.G. Kreĭn and M.A. Rutman /М.Г.Крейн, М.А.Рутман/
Линейные операторы, оставляющие инвариантным конус в пространстве Банаха, Успехи мат. наук 3, № 1 /1948/, 3 - 95.

[26] M.G. Kreĭn and Ju.L. Šmul'jan /М.Г.Крейн, Ю.Л.Шмульян/

a) О плюс-операторах в пространстве с индефинитной метрикой, Мат. исследования 1, вып. 1 /1966/, 131 - 161.
Engl. transl.: Amer. Math. Soc. Transl. (2) 85 (1969), 93 - 113.

b) $\mathfrak{J}$-полярное представление плюс-операторов, Мат. исследования 1, вып. 2 /1966/, 172 - 210.
Engl. transl.: Amer. Math. Soc. Transl. (2) 85 (1969), 115 - 143.

[27] R. Kühne
Über eine Klasse J-selbstadjungierter Operatoren, Math. Ann. 154 (1964), 56 - 69.

[28] H. Langer

a) Zur Spektraltheorie J-selbstadjungierter Operatoren, Math. Ann. 146 (1962), 60 - 85.

b) Eine Verallgemeinerung eines Satzes von L.S. Pontrjagin, Math. Ann. 152 (1963), 434 - 436.

c) Замечание об инвариантных подпространствах линейных операторов в банаховых пространствах с индефинитной метрикой, Мат. исследования 4, вып. 1 /1969/, 27 - 34.

d) Spektraltheorie J-selbstadjungierter Operatoren und einige Anwendungen auf die Schar $\lambda^2 I + \lambda B + C$, Habilitationsschrift, Techn. Univ. Dresden 1965.

[29] **H. Langer, K. Daho**
Sturm-Liouville operators with an indefinite weight function, Proc. Roy. Soc. Edinburgh Sect. A 78 (1977), 161 - 191.

[30] **L.A. Ljusternik and V.I. Sobolev** /Л.А.Люстерник, В.И.Соболев/
Элементы функционального анализа, ГИТТЛ, Москва, I95I.
Engl. transl.: Ungar, New York, 1961.

[31] **A.I. Mal'cev** /А.И.Мальцев/
Основы линейной алгебры, Наука, Москва, I970.
Engl. transl. of 2-nd (1956) ed.: Freeman, San Francisco, 1963.

[32] **M.A. Naĭmark** /М.А.Наймарк/
Аналог теоремы Стоуна в пространстве с индефинитной метрикой, Докл. АН СССР I70, № 6 /I966/, I259 - I26I.
Engl. transl.: Soviet Math. Dokl. 7 (1966), 1366 - 1368.

[33] **M.A. Naĭmark, A. I. Loginov and V.S. Šul'man** /М.А.Наймарк, А.И.Логинов, В.С.Шульман/
Несамосопряжённые алгебры операторов в гильбертовом пространстве, Итоги науки и техники, серия матем. анализ, том I2 /I974/, 4I3 - 465.
Engl. transl.: J. Soviet Math. 5 (1976), 250 - 278.

[34] **R.S. Phillips**
a) **Dissipative operators and hyperbolic systems of partial differential equations, Trans. Amer. Math. Soc. 90 (1959), 193 - 254.**
b) **On symplectic mappings of contraction operators, Studia Math. 3 (1968), 15 - 27.**

[35] **L.S. Pontrjagin** /Л.С.Понтрягин/
Эрмитовы операторы в пространствах с индефинитной метрикой, Изв. АН СССР Сер. мат. 8 /I944/, 243 - 280.

[36] **V.P. Potapov** /В.П.Потапов/
Мультипликативная структура $\mathfrak{J}$-нерастягивающих матриц-функций, Тр. Моск. мат. об-ва 4 /I955/, I25 - 236.
Engl. transl.: Amer. Math. Soc. Transl. (2), 15 (1960), 131 - 243.

[37] **F. Riesz, B. Sz.-Nagy**
Lecons d'analyse fonctionelle, 4th ed., Akad. Kiado, Budapest, Gauthiers Villars, Paris, 1965.
Engl. transl. of 2-nd ed.: Ungar, New York, 1955.

[38] **S.L. Sobolev** /С.Л.Соболев/
О движении симметрического волчка с полостью, наполненной жидкостью, Ж. прикл. механ. и техн. физ. 3 /I960/, 20 - 55.